A First Course in "In Silico Medicine"

Volume 3

Series Editor
Masao Tanaka
Professor of Osaka University

1-3 Machikaneyama, Toyonaka
Osaka 560-8531, Japan
tanaka@me.es.osaka-u.ac.jp

For further volumes:
http://www.springer.com/series/8773

Masao Tanaka · Shigeo Wada
Masanori Nakamura

Computational Biomechanics

Theoretical Background
and Biological/Biomedical Problems

 Springer

Masao Tanaka
Professor
Graduate School of Engineering Science
Osaka University
Toyonaka, Osaka, Japan
tanaka@me.es.osaka-u.ac.jp

Shigeo Wada
Professor
Graduate School of Engineering Science
Osaka University
Toyonaka, Osaka, Japan
shigeo@me.es.osaka-u.ac.jp

Masanori Nakamura
Associate Professor
Graduate School of Science and Engineering
Saitama University
Saitama, Saitama, Japan
masanorin@mech.saitama-u.ac.jp

ISBN 978-4-431-54072-4   e-ISBN 978-4-431-54073-1
DOI 10.1007/978-4-431-54073-1
Springer Tokyo Dordrecht Heidelberg London New York

Library of Congress Control Number: 2012933969

Printed on acid-free paper

Springer is part of Springer Science+Business Media (www.springer.com)

Preface

Computational biology and system biology are bases for *in silico* medicine, which consists of two main activities. The first is the development of mathematical models describing structures and functions of biological systems across a multiple scale augmented with experimental data from wet and dry biological and medical measurements. The second is the development of a simulator of biological functions in physiological and pathological situations to provide us with the behavior of components at multiple scales in a quantitative manner and to enable us to apply that information to medical problems. This book is the third volume of the textbook series A First Course in "In Silico Medicine". The first volume gives an introduction to computational physiology, and the second volume is devoted to computational electrophysiology. The main physical quantities discussed in these two volumes are electric/ionic currents and potentials related to electrical phenomena in biology. Other important physical quantities in biological systems are forces and deformations studied in mechanics, and these are the topics of this third volume.

Biomechanics is an area that deals with mechanical aspects, especially structures and functions, of hardware and software constructing biological systems of a living body. It is considered a relatively new area compared to mechanics, biology, and physiology, in spite of its long history at least from the ancient Greeks with Aristotle's treatises. In these decades, the area covered by biomechanics has expanded tremendously in accordance with rapid developments in life science. In fact, the dynamics of biomolecules such as DNA and proteins, mechanics of the whole cell and an intercellular structure, and so on are recent major subjects of biomechanics at a smaller scale, in addition to the advanced subjects of tissue and organ biomechanics of the musculo-skeletal system, the cardiovascular system, and other systems at a relatively larger scale. Qualitative advancement is another important direction in biomechanics because of rapid developments in biological/ biomedical measurement and imaging technologies as well as development of computer analysis and information technologies. The quantity of data nowadays is so huge that it results in a qualitative change in the meaning of data as a whole.

The quantitative increase in data, such as structural geometry obtained with CT and MR images, enables us to deal with a target body entirely as it is, and invites us into a virtual computer world that represents realistic biological tissue and/or organ structures, while classical studies focus on essential phenomena by simplification necessitated by the limited availability of fundamental data and solvable scale-problems. This direction is common in biomedical imaging supported by computer graphics and virtual reality technologies. Computational biomechanics lies in the same direction towards computational biology and medicine. The essences of computational biomechanics are mathematical and computational modeling of phenomena concerning deformations and forces.

In this textbook, mathematical fundamentals are limited to continuum mechanics, and no attention is paid to statistical, quantum, and relativistic mechanics. Therefore, a biological body is assumed to be a continuum body, because the continuum assumption is reasonable in many scales of size and time for biomedical problems except for molecule size, although any body is composed of atoms. Computational modeling is thus for discrete representation of a continuum within the scale of continuum mechanics. This is not limited to the geometrical modeling of an object body but also involves the mechanical modeling of phenomena under consideration. The degree of discretization is determined of course based on the objective of computational analysis in general, and it governs the adequacy of results of analyses. These are sometimes limited by the resolution of fundamental data available for analysis. These days, there are many powerful and function-rich engineering tools available in mechanics analysis on the software market. These are applicable to biomechanics problems as well, but this does not mean that computational biomechanics analysis is ready for every researcher or student. To be a reasonable user and become a smart user, it is essential to understand what the software tools are able to do and are doing for your problem. Biomechanics problems need mechanical modeling for biological bodies and environments, which are essentially different from common problems in mechanical, civil, and other engineering fields. For these reasons, the theoretical bases and assumptions should be understood and therefore this textbook is self-contained, covering both the basics of continuum mechanics of biosolids and biofluids and the theoretical core of computational methods for continuum mechanics analyses. Several biomechanics problems in orthopedic and cardiovascular biomechanics and other areas are provided for better understanding of computational analysis background and modeling issues in biomechanics problems. These are mainly direct analyses but include back or inverse analyses as well.

A standard direct analysis is the first key of computational biomechanics in *in silico* medicine that gives us much quantitative information for biological and biomedical phenomena in practical problems in medicine. A back/inverse analysis extended from a direct one is the promising second key as expected for the model-based prediction towards *in silico* medicine. We hope readers will be interested in the further development of biomechanics and contribution to predictive medicine.

We are grateful for the support of the Japan MEXT program "in silico medicine" at Osaka University, and we thank the staff at Springer Japan for their patience and encouragement.

Masao Tanaka
Shigeo Wada
Masanori Nakamura

Contents

Chapter 1
Introduction

Biomechanics is a branch of mechanics that studies living systems with mechanical disciplines. The bodies composing living system are macroscopically understood as continuum media in general, although they have hierarchical structure from biological molecule, cell to varieties of hard/soft tissues forming organs of individual living body. This chapter provides a brief introduction of biomechanics of biosolids and biofluids in continuum level using one-dimensional system. It is also given as a preparation for the following chapters how the biomechanical continua are treated in experimental, theoretical and computational methods of mechanics.

Keywords Mechanics in living system • One-dimensional biofluid mechanics • One-dimensional biosolid mechanics

1.1 Biomechanics: Mechanics in/for Biology and Medicine

Mechanics is an area of physics and studies the movement and deformation of physical objects under applied forces and displacements. In the classical mechanics, objective bodies are a system of a single particle/set of particles, of a rigid and deformable solid body, and/or of a fluid of liquid and/or gas. It works as a main scientific basis of mechanical engineering that is a key branch of engineering in industrial civilization. Physical objects generally include living systems and non-living systems. In modern history of mechanics and mechanical engineering, major efforts have been made for physical objects of nonliving systems intensively, and limited attentions have been paid for physical objects of living systems which are targeted objectives in biology and medicine. In the long history of development of mechanical science, however, scientists in mechanics investigated living systems as well as nonliving systems. Followings are just a couple of instances among many. Hooke's law is a law of elasticity in solid mechanics

M. Tanaka et al., *Computational Biomechanics*, A First Course in "In Silico Medicine" 3, DOI 10.1007/978-4-431-54073-1_1, © Springer 2012

named after the British physicist Robert Hooke (1635–1703) who is known in biology as the nomenclator of cell, the functional basic unit of biological organism. The French physicist Jean Louis Marie Poiseuille formulated the Poiseuille law known as Hagen–Poiseuille law in these days, a physical law in fluid mechanics describing the pressure drop in a laminar viscous and incompressible fluid flow in a cylindrical tube. The equation was successfully applied for the physiological system of the blood flow in capillaries. There were many others interested in mechanics of living systems.

The definition of biomechanics is not straightforward since it is rather a new branch in modern mechanics and the scope of biomechanics has been spreading. Some of them are quoted in the below: Hatze (1974) described "Biomechanics is the study of the structure and function of biological systems by means of the methods of mechanics." Yuan-Cheng Fung stated "Biomechanics is mechanics applied to biology" in Fung (1981) and "Biomechanics aims to explain the mechanic of life and living. From molecules to organisms, everything must obey the laws of mechanics" in Fung (1990). Humphrey and Delange (2004) considered "Biomechanics is the development, extension, and application of mechanics for the purposes of understanding better the influence of mechanical loads on the structure, properties and function of living things."

The rigorous definition is important from the viewpoint of science. Meanwhile, it is necessary to adjust the standpoint flexibly for the real problems under consideration, and the definition is not mentioned more and steps back to a generous but a bit vague standpoint for the realistic biomechanical aspect emerging in biomedical engineering. The prefix "bio" relating to living things suggests roughly and generally that biomechanics is a branch of mechanics studying living systems by means of methods of mechanics or mechanical science, as is similar to biophysics studying biological systems by using the methods of physical science.

1.2 One-Dimensional Mechanics of Biosolids and Biofluids

1.2.1 A Dip into Biosolid Mechanics

One-dimensional system will be a point of start for the general three-dimensional systems. A bone block of length L_0 and cross-section area A_0 of square D_0 on a side shown in Fig. 1.1 is the first biosolid system considered here. The attention is focused on the force and elongation along longitudinal direction. The one end of block was fixed and the axial force F is applied to the other end in the tensile direction. The displacement u of the force-applied end is the elongation of the block and a measure of the deformation. This is dependent on the magnitude of force applied, and is represented as the force-elongation (or displacement) relationship (Fig. 1.2a). This force-elongation relationship of course depends on the size of block, and is a structural property of the bone block but not a material property of the bone.

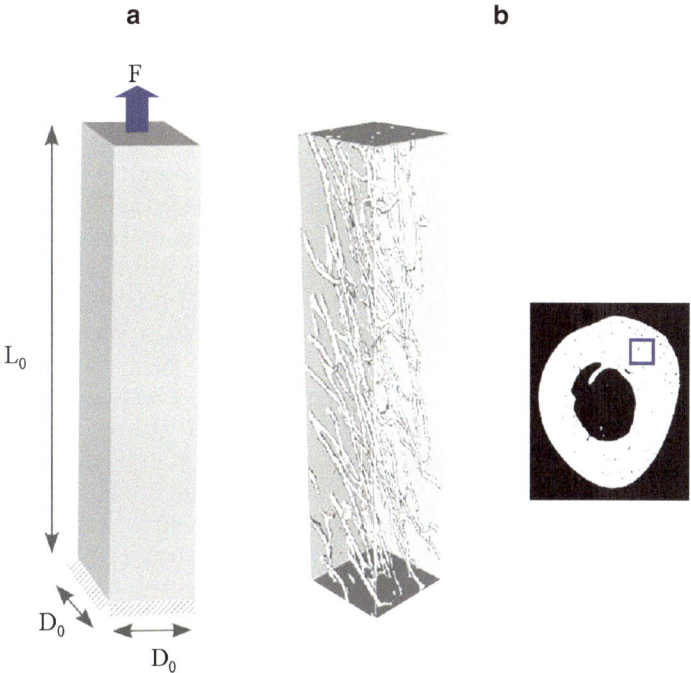

Fig. 1.1 Block specimen of bone. (**a**) Specimen size and (**b**) canal network in cortical bone

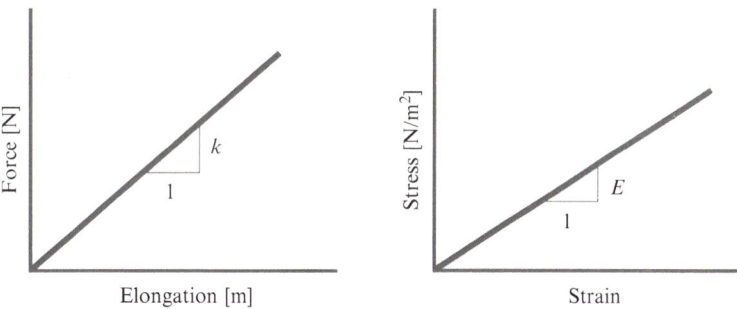

Fig. 1.2 Load-deformation diagram of linear elasticity. *Left*: force–elongation relationship as structure and *Right*: stress–strain relationship as material

To eliminate the influence of the block size and to reach the material property of bone, stress and strain are used instead of force and elongation. Mathematically, they are given as

$$\sigma = \frac{F}{A_0} \, , \, \varepsilon = \frac{u}{L_0} . \tag{1.1}$$

Note that stress σ is a force per unit area and strain ε is an elongation per unit length. The stress–strain relationship is reasonably written as a linear function as

$$\sigma = E\varepsilon \qquad (1.2)$$

with slope E representing the elastic modulus of the bone (Fig. 1.2b). This is known as Hooke's law, and this elastic modulus E is named as Young's modulus. Hooke's law is the fundamental characteristics of a linear elastic material. In SI system, the unit of force is a newton ($\mathrm{N} = \mathrm{kg\ m/s^2}$), the unit of length is a meter (m), and the unit of stress is thus a pascal ($\mathrm{Pa} = \mathrm{N/m^2}$) It is noted that strain is dimensionless and has no unit. The unit of the elastic modulus is also a pascal. The stress and strain will be expected to be the same at any points in the bone block, when the bone block is of homogeneous. Figure 1.1b is the micro CT image of a bone block taken from cortical shaft of murine tibio-fibula bone. As seen, it has many pores of canal network and is not homogeneous in detail. The stress/strain and the elastic modulus, nevertheless, give us important information on macroscopic biomechanical characteristics of bone as a homogeneous continuum. As the whole bone block, the force–elongation relationship is then represented by using the elastic modulus

$$\frac{F}{A_0} = E\frac{u}{L_0} \ \text{or}\ F = ku \qquad (1.3)$$

where $k = \frac{EA_0}{L_0}$ stands for the spring constant of the bone block with a specific block size. It is noted that the cross-sectional area decreases in accordance with the increase of longitudinal elongation, and the true stress at the elongated state of length $L = L_0 + u$ should be defined by using the deformed cross-sectional area A. In the case of bone, the strain is small within the physiological range of stress, and a change in the cross-sectional area is also small. Thus, the effect of cross-sectional change can be ignored generally in stress evaluation.

For the similar one-dimensional system of soft tissue of tendon/ligament, blood vessels, and others, the force–elongation relationship is different from the straight line observed for bone (Fig. 1.3). Stress and strain by (1.1) works but some extensions are needed. The linear stress–strain relationship in (1.2) is not applicable, and a nonlinear function is expected. The soft tissue shows larger deformation than bone, a typical hard tissue. In this situation, defined are Lagrange (or nominal) stress t, Cauchy (or true) stress σ and Kirchhoff stress s, and corresponding nominal strain e, logarithmic strain ε and Green's strain E as follows.[1]

---

[1] The stress σ in (1.1) is a nominal stress, though the symbol is same as the true stress of (1.5). The three stresses t, σ and s introduced here are approximately identical for the infinitesimal strain, where the three strains introduced here are approximately the same each other.

Fig. 1.3 Load-deformation
diagram of nonlinear
elasticity

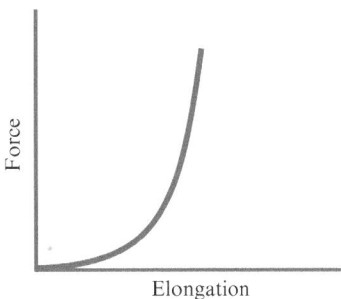

Force

Elongation

$$t = \frac{F}{A_0} \;,\; e = \int_{L_0}^{L} \frac{dL}{L_0} = \frac{L - L_0}{L_0} = \frac{L}{L_0} - 1 = \frac{u}{L_0} \;, \tag{1.4}$$

$$\sigma = \frac{F}{A} \;,\; \varepsilon = \int_{L_0}^{L} \frac{dL}{L} = \ln \frac{L}{L_0} \;, \tag{1.5}$$

$$s = \frac{F}{A_0} \frac{L_0}{L} \;,\; E = \frac{1}{2} \frac{L^2 - L_0^2}{L_0^2} = \frac{1}{2} \left[\left(\frac{L}{L_0} \right)^2 - 1 \right]. \tag{1.6}$$

The stresses and strains by (1.4) and (1.6) are defined for the reference configuration without deformation, and those by (1.5) are for the deformed configuration. The work in unit volume done by a small increase in strain is represented by these pairs of stress and strain like as

$$dW_0 = \frac{F}{A_0} \frac{dL}{L_0} = tde = sdE \tag{1.7}$$

for volume $V_0 = A_0 L_0$ in reference configuration or

$$dW = \frac{F}{A} \frac{dL}{L} = \sigma d\varepsilon \tag{1.8}$$

for volume $V = AL$ in deformed configuration. Conservation of mass with densities ρ_0 and ρ for V_0 and V gives

$$\rho dW_0 = \rho_o dW \tag{1.9}$$

and the relations between three stresses are derived:

$$t = s \frac{L}{L_0} \;,\; \sigma = t \frac{\rho}{\rho_0} \frac{L}{L_0} \;,\; s = \sigma \frac{\rho_0}{\rho} \left(\frac{L_0}{L} \right)^2. \tag{1.10}$$

The stress–strain relationship can be described using any pair of stress and strain. The pair of nominal stress and nominal strain is a conventional choice in the uniaxial tensile/compression experiment. Many varieties of a nonlinear function are recruited for describing the stress–strain relationship of soft tissue. Some classes of soft tissue are known to be well represented by an exponential function of

$$t = a\left\{\exp^{b(\lambda-1)} - 1\right\}$$ (1.11)

which implicitly states that the stress t is linearly correlated with the slope $\frac{dt}{d\lambda}$ with respect to the stretch $\lambda \equiv \frac{L}{L_0} = 1 + e$:

$$\frac{dt}{d\lambda} = b(t + a).$$ (1.12)

Many others are represented by a power function and a logarithm function.

Further on biosolid mechanics and its computational methods for in silico medicine will be found in Chap. 2.

Exercise 1.1 Examine the effect of magnitude of stretch ratio λ on the values of strains e, ε and E defined by (1.4)–(1.6). For instance, evaluate the difference among them for $\lambda = 1.001$ and $\lambda = 1.5$, that is the cases of 0.1% and 50% nominal strains.

1.2.2 A Dip into Biofluid Mechanics

An air flow in an individual airway tube of lung and a blood flow in an individual blood vessel will be most basic biofluid systems of one dimension. Air and blood are viscous fluid both, although the viscosity of air is smaller than that of blood. Viscosity is a resistance caused by shear deformation of a fluid and is defined as the coefficient between the shear stress τ and the velocity gradient $\frac{du}{dy}$ in direction y perpendicular to the direction of flow u. Newton's law of viscosity is written as

$$\tau = \mu \frac{du}{dy}$$ (1.13)

and μ is the coefficient named viscosity having a unit of [Pa s] (Fig. 1.4). The fluid is called Newtonian fluid when viscosity is constant.

The simplest mechanics of tube flow is of an incompressible Newtonian fluid. Suppose a steady, uniaxial, axisymmetric laminar flow in an individual vessel or airway of straight tube of diameter $2a$. In this situation, non-zero component of velocity is only an axial component, u, which is expressed as a function of radial position r. The equilibrium of the pressure drop over the section $\frac{dP}{dx}$ and the shear stress τ acting on the perimeter of the section at the tube wall is described by the Stokes equation;

$$\tau = -\frac{r}{2}\frac{dP}{dx}$$ (1.14)

and the law of viscosity in radial direction is

Fig. 1.4 Newton's law
of viscosity

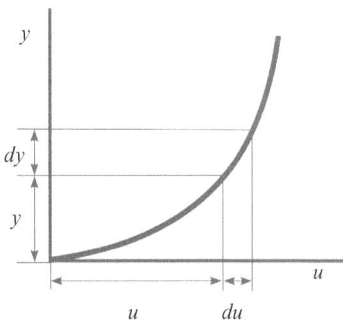

Fig. 1.5 Velocity profile
of Hagen–Poiseuille flow

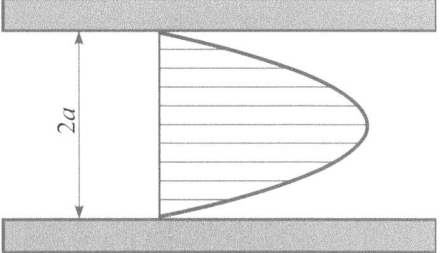

$$\tau = -\mu \frac{du}{dr}. \tag{1.15}$$

Under the non-slip condition $u = 0$ on the wall, a combination of (1.14) and (1.15)
yield a velocity profile within a tube;

$$u = -\frac{1}{4\mu}(a^2 - r^2)\frac{dP}{dx} \tag{1.16}$$

which describes that the flow takes a parabolic velocity profile, as is illustrated in
Fig. 1.5. Hagen–Poiseuille law is then derived as

$$Q = -\frac{\pi a^4}{8\mu}\frac{dP}{dx} \tag{1.17}$$

which relates the pressure gradient $\frac{dp}{dx}$ to the volumetric flow rate Q. A cross-
sectionally mean velocity U is gained by dividing (1.17) with the cross-sectional
area πa^2;

$$U = -\frac{a^2}{8\mu}\frac{dP}{dx}. \tag{1.18}$$

Fig. 1.6 Velocity profile of
Casson flow

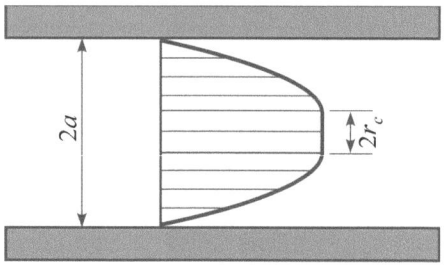

The assumption of Newton's law of viscosity is reasonable for the airflow in
lung airway, but the incompressible assumption is not in general for airflow. Blood
is a red cell-rich fluid, and its viscosity is dependent on the red cell contents and the
shear velocity. Thus, the blood is essentially a non-Newtonian fluid. In spite of
these facts, Poiseuille flow is often used for the first order estimation of mechanical
behavior of these cases because of its theoretical completeness.

Casson fluid is an improved expression of viscosity of blood. It is given by

$$\tau^{\frac{1}{2}} = \tau_C^{\frac{1}{2}} + (\eta \dot{\gamma})^{\frac{1}{2}} \tag{1.19}$$

where τ_C is the yield stress in shear and η is called Casson's coefficient of viscosity.
This describes the behavior of blood for a small shear strain rate well. In the case of
tube flow, the shear stress τ is proportional to the radial position as is in Stokes
equation (1.14), and the shear strain rate $\dot{\gamma} = -\dfrac{du}{dr}$ is zero, i.e. u is constant for the
core flow region of $r \leq r_C$ of $\tau_C = -\dfrac{r_C}{2}\dfrac{dP}{dx}$. Casson's equation (1.18) holds for
the region $r_C \leq r \leq a$ between the core flow and the wall. Figure 1.6 illustrates the
velocity profile of Casson flow. The flow rate of a Casson flow is expressed as a
modified formula of the Poiseuille flow (1.17);

$$Q = -\frac{\pi a^4}{8\mu}\frac{dp}{dx}F(\xi) \tag{1.20a}$$

with

$$F(\xi) = 1 - \frac{16}{7}\xi^{1/2} + \frac{4}{3}\xi - \frac{1}{21}\xi^4 \tag{1.20b}$$

$$\xi = \left(\frac{2\tau_c}{a}\right)\left(-\frac{dp}{dx}\right)^{-1}. \tag{1.20c}$$

We assumed a rigid wall of a flow channel thus far. In fact, blood vessels and
airways are comprised of soft tissues, whereby often necessitating consideration of

wall elasticity in the formulation of flow behaviors. For such a situation, we need to include the relationship between transmural pressure p (pressure difference between in- and out-sides of the channel) and cross-sectional area of the flow channel A. A rigorous approach to seek for the p–A relation is quite challenging. However, we make following assumptions to simplify the problem: (1) wall viscoelasticity is negligible, (2) the effect of dA at x_1 on A at x_2 is negligible and (3) the outside pressure is constant. If a tube radius a is assumed to be linearly correlated with pressure p, the Hagen–Poiseuille law (1.17) is rewritten as

$$Q = -\frac{\pi a^4}{4\mu\alpha}\frac{da}{dx}. \tag{1.21}$$

For further modeling of an inviscid flow in an elastic tube, we use equations of conservations of mass and momentum. In one-dimensional analysis, they are described with

$$\frac{\partial A}{\partial t} + \frac{\partial}{\partial x}(uA) + pG = 0 \tag{1.22}$$

$$\frac{\partial u}{\partial t} + u\frac{\partial u}{\partial x} + \frac{1}{\rho}\frac{\partial p}{\partial x} + \frac{R}{\rho}Au = 0 \tag{1.23}$$

where u is a cross-sectionally mean flow velocity $u(x, t)$, and G and R are leakage from the wall and resistance. In addition, we introduce the equation for wall motion as

$$p = P(A) \tag{1.24}$$

which simply states that the pressure is a function of the cross-sectional area A. We also introduce wall distensibility as

$$D = \frac{1}{A}\frac{dA}{dp} = \frac{1}{AP'(A)} \tag{1.25}$$

and the velocity of wave propagation c as

$$c^2 = \frac{1}{\rho D} = \frac{A}{\rho}P'(A). \tag{1.26}$$

In order to linearise (1.23), we make an extra assumption that the velocity of wave propagation is much larger than that of fluid at all times. This assumption is reasonable for human physiological systems, since the minimum wave propagation velocity is 4 m/s in the ascending aorta whereas the maximum flow velocity is 1 m/s (also in the ascending aorta). If G and R are zero and a radial change of the tube in the x-direction is negligibly small in comparison to that in time, we can omit the second term of (1.23), $u\frac{\partial u}{\partial x}$. Using (1.22), (1.24) and (1.26) with some mathematical operations, we obtain

$$\frac{\partial p}{\partial t} + \rho c^2\frac{\partial u}{\partial x} = 0. \tag{1.27}$$

Similarly, we linearise (1.23) by assuming that $\frac{\partial u}{\partial t} \gg u \frac{\partial u}{\partial x}$ and gain

$$\frac{\partial u}{\partial t} + \frac{1}{\rho} \frac{\partial p}{\partial x} = 0. \tag{1.28}$$

Then, differentiating (1.27) with respect to t, and (1.28) with respect to x, we get

$$\frac{\partial^2 p}{\partial x^2} - \frac{1}{c^2} \frac{\partial^2 p}{\partial t^2} = 0 \tag{1.29}$$

This is called the wave equation describing propagation of the pressure through the flow channel at speed $\pm\, c$.

One-dimensional mechanics of biofluid has many limitations to describe the real phenomenon. However, it gives us the first order approximation and the basis for three-dimensional study given in Chap. 3.

Exercise 1.2 Describe the velocity profile of Casson tube flow. That is, first, substitute (1.14) and (1.15) into (1.19), solve the obtained equation with respect to $\frac{du}{dr}$, and finally integrate it under the non-slip condition at the wall.

Exercise 1.3 Derive the wave equation (1.27) using (1.22), (1.24) and (1.26).

Exercise 1.4 Derive the wave equation (1.29) using (1.27) and (1.28).

1.3 Addendum to One-Dimensional Mechanics

1.3.1 Law of Mixture

Biological tissues are generally composed of multiple materials constructing multilevel hierarchical structure, although many of engineering materials are of monolithic or of combination of materials that are not distinguishable like an alloy. For example, tendons connecting muscle to bone and ligaments connecting bone to other bone are both fibrous connective tissues made of collagen fiber bundles. These tissues exhibit a composite structure of collagen fibers embedded in a matrix interconnecting them at different levels in scale. Due to one-dimensional geometry in shape, these tissues are used as a typical example of one-dimensional biomechanics introduced in Sect. 1.2.1, as a first order modeling.

A stress–strain relationship describes the mechanical property of tendon or ligament at the scale of whole tissue. It is also possible to explore the stress–strain relationship of collagen fiber or matrix as well as the tissue. These stress–strain relationships are different each other and also different from that as a whole tissue. The relation among these stress–strain relationships will be another topic of mechanics in one-dimension. This situation is similar to that of composite material such as fiber-reinforced materials used in industry.

As the simplest situation, let us consider a composite material made of matrix and fiber constituting a structure similar to that of tendon/ligament but exhibiting

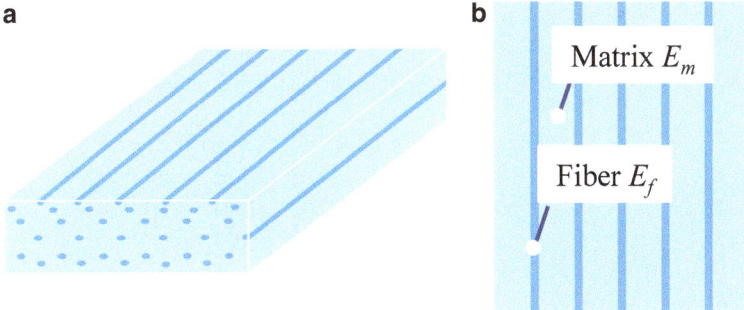

Fig. 1.7 Aligned continuous fiber composite

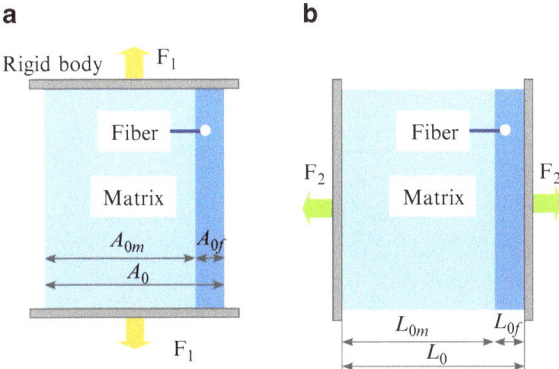

Fig. 1.8 Models for law of mixture. (**a**) Voigt model, (**b**) Reuss model

linear elastic behavior represented by (1.2) or (1.3) different from nonlinear behavior of real tendon/ligament, illustrated in Fig. 1.7. When the force is applied in the longitudinal direction of fiber orientation, the applied force F_1 is shared between the matrix F_m and the fiber F_f as

$$F_1 = F_m + F_f = A_{0m}\sigma_m + A_{0f}\sigma_f \tag{1.30}$$

where A_{0m} and A_{0f} are total cross-sectional areas of matrix and fiber materials illustrated in Fig. 1.8a. The linear stress–strain relationships of matrix and fiber

$$\sigma_m = E_m\varepsilon_m \text{ and } \sigma_f = E_f\varepsilon_f \tag{1.31}$$

yield the stress–strain relationship as the composite material as

$$\sigma_{c1} \equiv \frac{F_1}{A_0} = \left(\frac{A_{0m}}{A_0}E_m + \frac{A_{0f}}{A_0}E_f\right)\varepsilon_{c1} \tag{1.32}$$

Fig. 1.9 Elastic moduli by
law of mixture

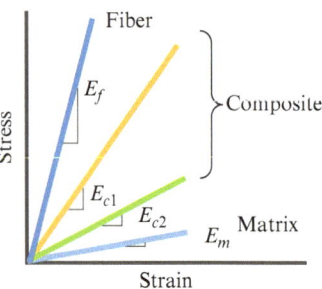

under the isostrain condition $\varepsilon_m = \varepsilon_f \equiv \varepsilon_{c1}$, that is equal strain in matrix and fiber
assuming perfect bonding between matrix and fiber materials, where $A_0 = A_{0m} + A_{0f}$ is the cross-sectional area as the composite and σ_{c1} denotes the stress as the
composite. This gives us the stress and elastic modulus in longitudinal direction as

$$\sigma_{c1} = V_m\sigma_m + V_f\sigma_f \quad \text{and} \quad E_{c1} = V_mE_m + V_fE_f \tag{1.33}$$

where $V_m = \frac{A_{0m}}{A_0} = (1 - V_f)$ and $V_f = \frac{A_{0f}}{A_0} = (1 - V_m)$ are volume fractions of
matrix and fiber material having the same length in longitudinal direction, respectively. This is the law of mixture for longitudinal direction.

When the force works in the direction perpendicular to the fiber orientation
(Fig. 1.8b), the elongation u is divided into elongations u_m and u_f of matrix and
fiber. The isostress condition $\sigma_m = \sigma_f \equiv \sigma_{c2}$ assuming the perfect bonding gives us
the law of mixture for transverse direction for strain and elastic modulus as follows:

$$\varepsilon_{c2} = V_m\varepsilon_m + V_f\varepsilon_f \quad \text{and} \quad \frac{1}{E_{c2}} = V_m\frac{1}{E_m} + V_f\frac{1}{E_f}. \tag{1.34}$$

The elastic moduli E_{c1} in longitudinal direction and E_{c2} in transverse direction as
the composite are different each other (Fig. 1.9). This means that the composite
have anisotropic mechanical property even when it is a composite of isotropic
materials, in which the elastic modulus is same in any direction.

The law of mixture is very useful as the first order analysis of the mechanical
property of composites. The concept does not require linear elasticity or anisotropy.
However, these and other assumptions such as continuous fiber and perfect bonding
are not always valid for biological tissues, and it is not possible to have the law of
mixture in analytical ways. Furthermore the tissue properties are not uniform and
region-dependent even in a single organ. Therefore we need another approach to
overcome this kind of difficulty in practice.

Exercise 1.5 Derive the law of mixture given in (1.34). Elongation is $u_c = u_m + u_f$
and strains of matrix and fiber are $\varepsilon_m = \frac{u_m}{l_{0m}}$ and $\varepsilon_f = \frac{u_f}{l_{0m}}$ where l_{0m} and l_{0f} denote the

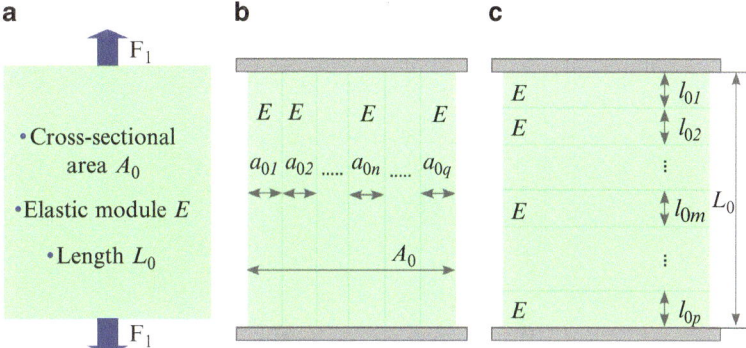

Fig. 1.10 Division of single continuum body. (**a**) Single continuum body, (**b**) parallel subdivision and (**c**) serial subdivision

equivalent lengths of matrix and fiber materials in transverse direction and the length as the whole composite is $l_0 = l_{0m} + l_{0f}$.

Exercise 1.6 Calculate the fiber volume fraction of a fiber composite based on the law of mixture, when the elastic modulus of as the composite is 700 MPa ($=700 \times 10^6$N/m^2) in longitudinal direction, the elastic modulus of matrix is 50 MPa and that of fiber is 1 GPa ($=10^9$N/m^2). Then, compare the elastic moduli of the composite in longitudinal and transverse directions.

1.3.2   Discrete Model of Continuum

Solid and fluid bodies considered in Sect. 1.2 are continuum. In the previous section for the law of mixture, the bodies are divided into parts of individual materials and the resultant force or elongation is divided accordingly as is in (1.30) or equation in Exercise 1.5. It is possible to divide single continuum body into a number of parts made of same material in parallel or serial, as is shown in Fig. 1.10. In parallel division, the cross-sectional area A_0 is represented as a collection of $a_{01}, a_{02}, \ldots$ and a_{0q}, and the applied force F is divided into q forces $F_1, F_2, \ldots, F_q$ as

$$F = F_1 + F_2 + \ldots + F_q. \tag{1.35}$$

Every division is made of the same material, and (1.3) defines spring constant k_n depending on each cross-sectional area of division n

$$F_n = k_n u_n \text{ and } k_n = \frac{E a_{0n}}{L_0} \ (n = 1,2,\ldots,q) \tag{1.36}$$

The perfect bonding condition between divisions, i.e. $u_1 = u_2 = \ldots = u_q = u$ results in the relation

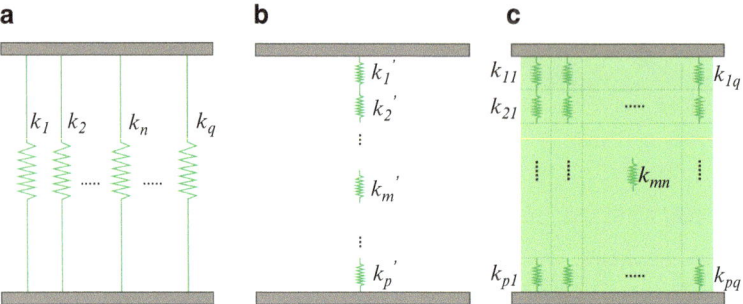

Fig. 1.11 Discrete models of single continuum body. (**a**) Parallel spring model, (**b**) serial spring model and (**c**) parallel and serial spring model

$$F = ku \ \text{and} \ k = k_1 + k_2 + \ldots + k_q. \tag{1.37}$$

The whole of cross-sectional area A_0 is equivalent to a parallel connection of springs with constant k_n corresponding to the cross-sectional area a_n (see Fig. 1.11a). This formulation is useful in the mechanical analysis of tissue exhibiting different elastic modulus through the cross-section. Individual spring constants are determined by (1.36) by choosing each of sufficiently small cross-sectional area so that the elastic modulus can be assumed to be constant. In the case of nonlinear material, the parallel division approach helps us to explore the force and elongation as the whole

$$F = F_1(u) + F_2(u) + \ldots + F_q(u) \tag{1.38}$$

Consider the serial division shown in Fig. 1.10c. In this case, the length L_0 is represented as a collection of $l_{01}, l_{02}, \ldots$ and l_{0p}, and the set of elongation u_m of each division is given by

$$u_m = \frac{F}{k'_m} \ \text{and} \ k'_m = \frac{EA_0}{l_{0m}} \ (m = 1, 2, \ldots, p). \tag{1.39}$$

Those help us formulate the total elongation $u = u_1 + u_2 + \ldots + u_p$ with the spring constant k'_m in accordance with the length division. Every division carries the same force F and the original body is represented as a serial connection of springs as is shown in Fig. 1.11b. This formulation is valid for the division for different elastic modulus as is in the section for law of mixture and also useful for nonlinear elastic cases as

$$u = u_1(F) + u_2(F) + \ldots + u_p(F). \tag{1.40}$$

More generally, the original body are divided in parallel and serial simultaneously as shown in Fig. 1.11c, and is represented as a parallel and serial connections of springs with constant

$$k_{mn} = \frac{Ea_{0n}}{l_{0m}} \ (m = 1,2, ..., p, \ n = 1,2, ..., q) \tag{1.41}$$

representing individual division of cross-sectional area a_{0n} and length l_{0m}. These give fundamentals for the discrete representation of continuum body of solids and fluids, although here explained are solid body examples only.

1.4   Biomechanics in Biology and Medicine: A Tiny Showcase

Young's modulus is the fundamental mechanical characteristics of solid body. Even for soft tissue exhibiting essentially nonlinear elastic characteristics, the tangential moduli are frequently used as the representative property. For example, the quadriceps tendon connects quadriceps muscle and patellar bone and patellar tendon (ligament) does the patellar bone and tibia. The quadriceps tendon and patellar ligament are serially connected between quadriceps muscle and tibia, but the mechanical properties of quadriceps tendon and patellar ligament are not same. These characteristics are quantitatively represented in terms of tangent modulus (Staubli et al. 1999) as shown in Fig. 1.12. Tangential moduli were evaluated at the load of 200 and 800 N corresponding to the toe part and linear part of force elongation curve. That is, the modulus of patellar ligament is significantly higher than of quadriceps tendon. This is a demonstrative example of the importance of biomechanics providing a quantitative measure characterizing biological solids so-called mechano-physiology.

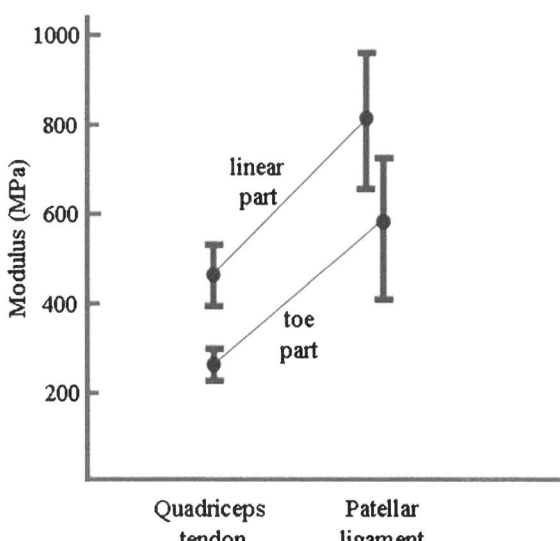

Fig. 1.12 Mechanical properties of quadriceps tendon and patellar ligament. *Source*: Taken from Staubli et al. (1999)

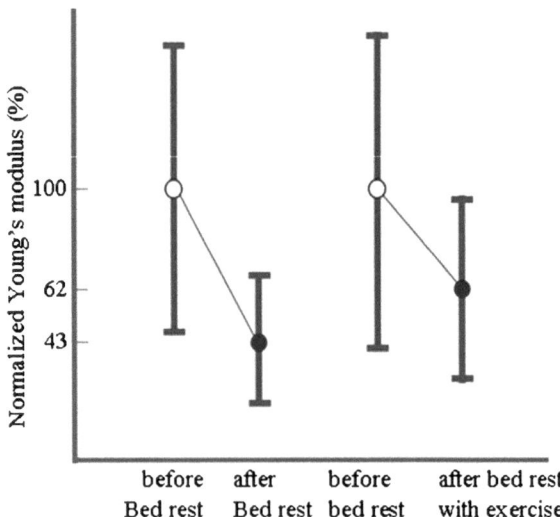

Fig. 1.13 Influence of microgravity environment on elastic modulus of human tendon. *Source*: Taken from Reeves et al. (2005)

Such characteristics of biological tissue are dependent on the biomechanical environment even for the specific site of a specific subject. For instance, the mechanical properties of human tendon are influenced by the gravity environment such as microgravity in space. Reeves et al. (2005) simulated the microgravity environment by 90 days of bed rest, and reported that Young's modulus of Gastrocnemius tendon decreased by 57% from 266.3 ± 137.5 MPa[2] to 113.6 ± 62.4 MPa after the bed rest and that the decrease of Young's modulus by bed rest was suppressed to 38% from 303.4 ± 150.8 MPa to 187.2 ± 100.5 MPa due to the exercise of the leg press for the hip, knee and ankle extensors and calf raise for the ankle extensors at every third day (Fig. 1.13). The load deformation data was nonlinear of course and Young's modulus is the average modulus for the force interval from 250 to 500 N. This is a typical example of biomechanical basics in space medicine demonstrating the importance of the mechanical environment on biological tissue and of the biomechanics providing a quantitative measure characterizing biosolids for mechano-biology and biomedicine.

The fluid mechanical discipline has also been applied to medicine. Pulsatility of blood flow due to heart beating causes oscillation of an arterial wall in the radial direction. This oscillation propagates toward the peripheral blood vessels. Despite the extreme complexity of wave propagation in living blood vessels due to wave reflection and energy dissipation, the wave propagation through systemic circulations is studied to diagnose arterial stiffness. Young (1809) studied the pulse wave propagation in terms of traveling speed in an ideal elastic tube. His study was extended by Korteweg (1878) and Moens (1878) who formulated the velocity of pulse wave propagation (pulse wave velocity, PWV) in a long straight elastic tube. The equation is known as the Moens–Korteweg equation, given by

---

[2] The expression $m \pm s$ denotes the average m and the standard deviation s.

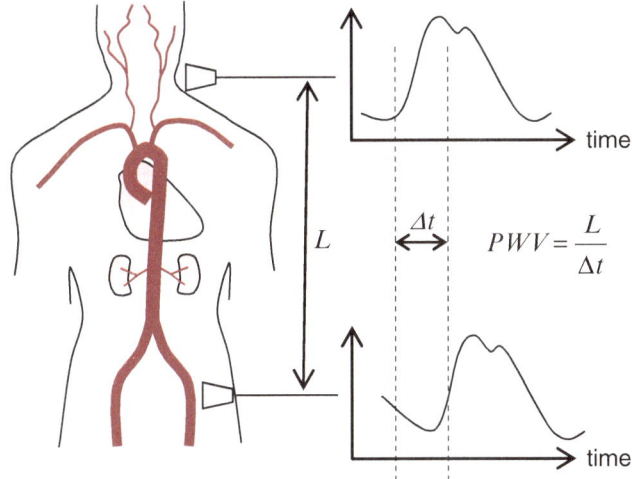

Fig. 1.14 Estimation of the PWV by the Frank method

$$PWV = \sqrt{\frac{Eh}{2\rho r_i}}, \tag{1.42}$$

where E is the Young's modulus of the wall, h is the wall thickness, ρ is the fluid density, and r_i is the internal radius of the tube. As seen in (1.42), the value of PWV becomes larger with an increase in the Young's modulus of the wall. Because of this simplicity, PWV is favored by medical doctors, and hence PWV is used as a measure of arterial stiffness. Figure 1.14 illustrates the measurement method of PWV (the Frank method) in clinical practice (Frank 1927). PWV is now used as an index of various systemic circulation diseases, including hypertension (Ting et al. 1991; Blacher et al. 1999), diabetes (Lehmann et al. 1992; Taniwaki et al. 1999), coronary artery disease (Triposkiadis et al. 1993), and eating disorders (Avolio et al. 1986; Hamazaki et al. 1988; Breithaupt-Grogler et al. 1997).

References

Avolio AP, Clyde KM, Beard TC, Cooke HM, Ho KK, O'Rourke MF (1986) Improved arterial distensibility in normotensive subjects on a low salt diet. Arteriosclerosis 6:166–169
Blacher J, Asmar R, Djane S, London GM, Safar ME (1999) Aortic pulse wave velocity as a marker of cardiovascular risk in hypertensive patients. Hypertension 33:1111–1117
Breithaupt-Grogler K, Ling M, Boudoulas H, Belz GG (1997) Protective effect of chronic garlic intake on elastic properties of aorta in the elderly. Circulation 96:2649–2655
Frank O (1927) Die Theorie der Pulswellen. Z Biol 85:91–130

Fung YC (1981) Biomechanics: mechanical properties of living tissues. Springer, New York, p 1

Fung YC (1990) Biomechanics: motion, flow, stress, and growth. Springer, New York, p.vii (preface)

Hamazaki T, Urakaze M, Sawazaki S, Yamazaki K, Taki H, Yano S (1988) Comparison of pulse wave velocity of the aorta between inhabitants of fishing and farming villages in Japan. Atherosclerosis 73:157–160

Hatze H (1974) The meaning of the term "biomechanics". J Biomech 7(2):189–190

Humphrey JD, Delange SL (2004) An introduction to biomechanics – solids and fluids, analysis and design. Springer, New York, p 3

Korteweg DJ (1878) Ueber die Fortpflanzungsgeschwindigkeit des Schalles in elastischen Röhren. Annalen der Physik 241:525–542

Lehmann ED, Gosling RG, Sonksen PH (1992) Arterial wall compliance in diabetes. Diabet Med 9:114–119

Moens AI (1878) Die Pulskurve. Brill, Leiden

Reeves ND, Maganaris CN, Ferretti G, Narici MV (2005) Influence of 90-day simulated microgravity on human tendon mechanical properties and the effect of resistive countermeasures. J Appl Physiol 98:2278–2286

Staubli HU, Schatzmann L, Brunner P, Rincon L, Nolte LP (1999) Mechanical tensile properties of quadriceps tendon and patellar ligament in young adults. Am J Sports Med 27(1):27–34

Taniwaki H, Kawagishi T, Emoto M, Shoji T, Kanda H, Maekawa K, Nishizawa Y, Morii H (1999) Correlation between the intima-media thickness of the carotid artery and aortic pulse-wave velocity in patients with type_2 diabetes. Vessel wall properties in type_2 diabetes. Diabetes Care 22:1851–1857

Ting CT, Chou CY, Chang MS, Wang SP, Chiang BN, Yin FC (1991) Arterial hemodynamics in human hypertension. Effects of adrenergic blockade. Circulation 84:1049–1057

Triposkiadis F, Kallikazaros I, Trikas A, Stefanadis C, Stratos C, Tsekoura D, Toutouzas P (1993) A comparative study of the effect of coronary artery disease on ascending and abdominal aorta distensibility and pulse wave velocity. Acta Cardiol 48:221–233

Young T (1809) On the functions of the heart and arteries. The Croonian Lecture. Phil Trans Roy Soc 99:1–31

Chapter 2
Mechanics of Biosolids and Computational Analysis

Behaviours of solid biological bodies are described in the context of solid mechanics. As is glanced in Chap. 1, load and deformation are characterized by stress and strain. They are introduced for one-dimensional mechanics in Sect. 1.2. First, this chapter gives their extension to three-dimensional mechanics under a small strain theory and a finite strain theory. It is followed by description of constitutive equations of linear elastic body and nonlinear hyperelastic bodies. Although biosolids behave as a viscoelastic body in general, the linear elasticity is a reasonably accounts for their behaviour as far as the deformation is small. Especially, the linear elastic theory works for hard biosolids such as bones in a physiological daily activity range. The linear elasticity is also useful for the first step of biomechanical analyses within the range of small deformation. For the finite deformations, a cyclic response in the physiological range is characterized by the concept of pseudo-elasticity with hyperelasticity. Second, constitutive equations are demonstrated for cortical bone and cancellous bone as a linear elastic body and for several soft tissues of an arterial wall, a skin and a cornea as typical examples of a nonlinear hyperelastic body. Third, an equilibrium problem is given as in the form of a virtual work and a stationary potential energy for linear elastic and nonlinear hyperelastic bodies. These provide us the basis for computational analyses of biosolids. Fourth, the fundamentals of a finite element method are given for a small strain theory and for a finite strain theory. Concept of the finite element approximation is explained for typical elements and the finite element equations are derived. Fifth, several computational biomechanics analyses are presented for orthopaedic, dental and ophthalmic biomechanics problems.

Keywords Biosolid mechanics • Stress and strain analyses • Finite element method • Hyperelastic soft tissue • Linear elastic hard tissue • Mechanical properties of biosolids

2.1 Fundamentals of Solid Mechanics

2.1.1 Stress and Force: Equilibrium Equations

In Sect. 1.2.1 for one-dimensional mechanics, the stress is introduced as the force per unit area and is assumed to be uniform over the cross section by paying attentions to one coordinate axis along the longitudinal direction only. Three-dimensional mechanics uses three coordinate axes, say x-, y- and z-axis or x_1-, x_2-, and x_3-axis for a Cartesian coordinate system. Accordingly, force has three components, and stresses intrinsically have components corresponding to each component of a force vector as $\frac{\Delta F_x}{\Delta A}$, $\frac{\Delta F_y}{\Delta A}$ and $\frac{\Delta F_z}{\Delta A}$ for an area ΔA. For the Cartesian coordinate system,[1] stresses are considered for three areas ΔA_x, ΔA_y and ΔA_z whose outward unit normal vectors are in the x, y, and z coordinate directions. Specifically, the stress components are defined as the limit of the ratios as

$$\sigma_{xx} = \lim_{\Delta A_x \to 0} \frac{\Delta F_x}{\Delta A_x} = \frac{dF_x}{dA_x}, \; \sigma_{yx} = \lim_{\Delta A_y \to 0} \frac{\Delta F_x}{\Delta A_y} = \frac{dF_x}{dA_y} \tag{2.1}$$

and so forth. Each of stress components has two indices: the first index is for the orientation of the area and the second one is for the direction of a force component. There are nine stress components in total

$$\left(\sigma_{ij} \right) = \begin{bmatrix} \sigma_{xx} & \sigma_{xy} & \sigma_{xz} \\ \sigma_{yx} & \sigma_{yy} & \sigma_{yz} \\ \sigma_{zx} & \sigma_{zy} & \sigma_{zz} \end{bmatrix} = \begin{bmatrix} \sigma_{11} & \sigma_{12} & \sigma_{13} \\ \sigma_{21} & \sigma_{22} & \sigma_{23} \\ \sigma_{31} & \sigma_{32} & \sigma_{33} \end{bmatrix} \tag{2.2}$$

as shown in Fig. 2.1 for an infinitesimal element, representing a point by a cube/hexahedron of an infinitesimal size. The components with the same first and second indices are normal stresses and the ones with different indices are shear stresses. The sign of stress component is positive when the sign of the orientation of area is the same as that of a working force. The value of each stress component is dependent on the choice of coordinate system,[2] although the physical stress condition is represented by the set of stress component σ_{ij} in any coordinate system.

The equilibrium among the stress components is considered for a hexahedron of $\Delta x_1 \times \Delta x_2 \times \Delta x_3$ shown in Fig. 2.2. When the stress at the centroid of the hexahedron is σ_{ij}, the stress components at the positive and negative faces perpendicular to

---

[1] Cartesian coordinate system is employed throughout this chapter unless otherwise noted. That is, all the components are Cartesian components.

[2] Stress components σ'_{ij} in an x'_i-coordinate system is calculated by using those σ_{ij} in an x_i-coordinate system as $\sigma'_{ij} = \sum_k \sum_l a_{ik} a_{jl} \sigma_{kl}$, where a_{ij} is components of the rotation matrix, or $\left(\sigma'_{ij} \right) = (a_{ik})(\sigma_{kl})(a_{jl})^T$ in matrix notation.

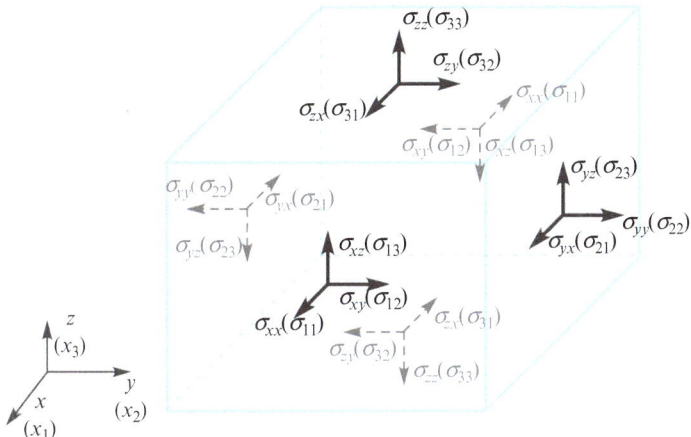

Fig. 2.1 Stress components

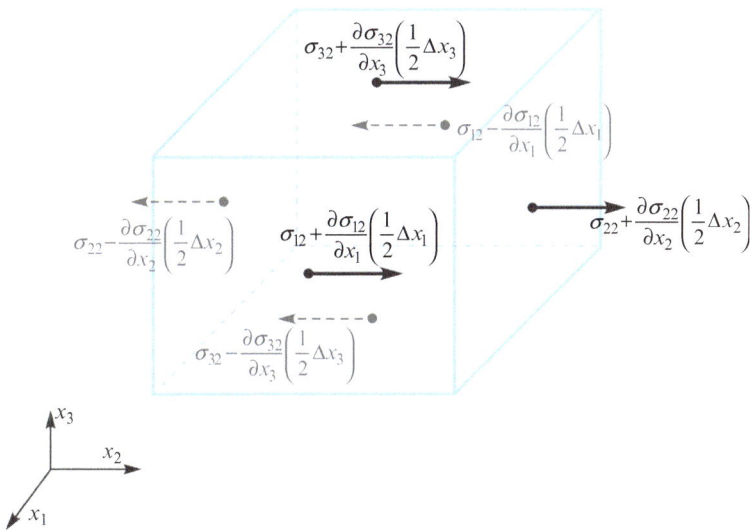

Fig. 2.2 Stress components in x_2 direction

x_i axis are $\sigma_{ij} + \frac{\partial \sigma_{ij}}{\partial x_i}\left(\frac{1}{2}\Delta x_i\right)$ and $\sigma_{ij} - \frac{\partial \sigma_{ij}}{\partial x_i}\left(\frac{1}{2}\Delta x_i\right)$ with neglecting the higher order terms. By taking the respective area for each stress component, the equilibrium of force in x_j-direction is then

$$\sum_i \frac{\partial \sigma_{ij}}{\partial x_i} = 0 \text{ for } j = 1, 2, 3. \tag{2.3}$$

Taking the body force b_i per unit volume into account, the equilibrium then becomes

$$\sum_i \frac{\partial \sigma_{ij}}{\partial x_i} + b_j = 0 \quad \text{for } j = 1, 2, 3. \tag{2.4}$$

or simply as

$$\frac{\partial \sigma_{ij}}{\partial x_i} + b_j = 0 \quad \text{for } j = 1, 2, 3 \tag{2.5}$$

by using the summation convention.[3] The summation convention is employed for index notation throughout this volume. Using the xyz axes notation, the equilibrium is further rewritten as

$$\frac{\partial \sigma_{xx}}{\partial x} + \frac{\partial \sigma_{yx}}{\partial y} + \frac{\partial \sigma_{zx}}{\partial z} + b_x = 0$$

$$\frac{\partial \sigma_{xy}}{\partial x} + \frac{\partial \sigma_{yy}}{\partial y} + \frac{\partial \sigma_{zy}}{\partial z} + b_y = 0$$

$$\frac{\partial \sigma_{xz}}{\partial x} + \frac{\partial \sigma_{yz}}{\partial y} + \frac{\partial \sigma_{zz}}{\partial z} + b_z = 0. \tag{2.6}$$

The equilibrium condition of moments among stress components about coordinate axes requests $\sigma_{ij} = \sigma_{ji}$ or $\sigma_{xy} = \sigma_{yx}$, which is the symmetry of the stresses in a matrix format of (2.2). Therefore, three normal and three shear stress components, that is, six components in total are independent in nine stress components defined.

Exercise 2.1 Examine the symmetric condition of stress components. The stress component represents a force per unit area. The force is, thus, the product of a stress component and the area of hexahedron face where it is working, and the moment is the product of force and the moment arm about the axis of rotation.

Exercise 2.2 Using the stress vector[4] of independent six stress components defined as $\boldsymbol{\sigma} = [\sigma_{xx} \ \sigma_{yy} \ \sigma_{zz} \ \sigma_{yz} \ \sigma_{zx} \ \sigma_{xy}]^T$, describe the equilibrium equations (2.6) in a matrix equation of $\partial \boldsymbol{\sigma} + \mathbf{b} = \mathbf{0}$. The matrix ∂ of 3×6 has differential operators as its components.

---

[3] Rules of summation convention are (1) each index can appear once or twice in any term and (2) every index appeared twice is summed over its range. For example, $a_{ii} = \sum_i a_{ii} = a_{11} + a_{22} + a_{33} = \sum_j a_{jj} = a_{jj}, a_{ij}b_j = \sum_j a_{ij}b_j = a_{i1}b_1 + a_{i2}b_2 + a_{i3}b_3$. Expression $a_{ij}b_j = c_j$ and $a_{ij}b_j + c_{ji}d_j$ are valid, but $a_{jj}b_j$ and $a_{ij}b_j = c_j$ are invalid.

[4] This is the Voigt notation in vector form.

Fig. 2.3 Line elements in undeformed and deformed configuration and displacement

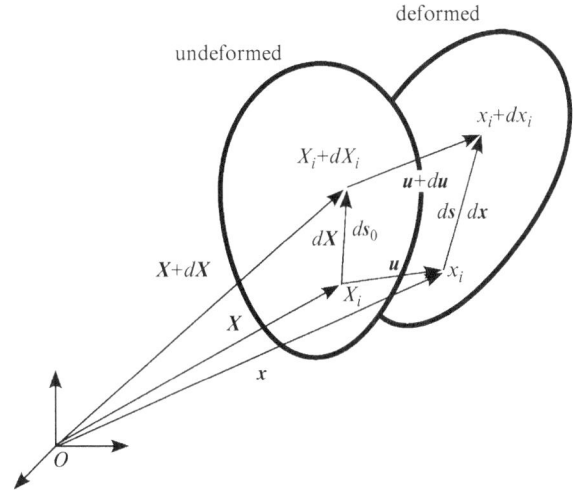

2.1.2   Strain and Displacement: Kinematic Equations

Deformable solid body subjected to forces transforms its configuration from undeformed reference one to deformed one. The deformation is described by using the displacement of a point of the body. The displacement vector **u** is defined as the difference of position vector **x** of a point in the deformed configuration from its position vector **X** in the undeformed configuration as shown in Fig. 2.3. It is written as

$$u_i = x_i(X_j) - X_i \ \text{ or } \ u_i = x_i - X_i(x_j). \tag{2.7}$$

Let us consider a line element between two points **X** and **X** + d**X** in the undeformed configuration. The length of line element ds_0, i.e. the distance between two points is

$$ds_0{}^2 = dX_i dX_i \ \text{ i.e. } ds_0{}^2 = dX_1{}^2 + dX_2{}^2 + dX_3{}^2 \tag{2.8}$$

and the distance ds between these points in the deformed configuration is

$$ds^2 = dx_i dx_i \ \text{ i.e. } ds^2 = dx_1{}^2 + dx_2{}^2 + dx_3{}^2. \tag{2.9}$$

These two distances are related as

$$ds^2 - ds_0{}^2 = 2E_{ij}dX_i dX_j = 2e_{ij}dx_i dx_j \tag{2.10}$$

by using Green strain tensor E_{ij} or using Almansi strain tensor e_{ij} defined as follows:

$$E_{ij} = \frac{1}{2}\left(\frac{\partial u_i}{\partial X_j} + \frac{\partial u_j}{\partial X_i} + \frac{\partial u_k}{\partial X_i}\frac{\partial u_k}{\partial X_j}\right) \tag{2.11}$$

$$e_{ij} = \frac{1}{2}\left(\frac{\partial u_i}{\partial x_j} + \frac{\partial u_j}{\partial x_i} + \frac{\partial u_k}{\partial x_i}\frac{\partial u_k}{\partial x_j}\right). \tag{2.12}$$

Green strain tensor E_{ij} and Almansi strain tensor e_{ij} are also referred to as Lagrangian strain tensor and Eulerian strain tensor, respectively. These give the kinematic equations of finite strain theory providing the nonlinear relation between the displacement and the strain.

For the case of small deformation, $\mathbf{x} \approx \mathbf{X}$ and $|\frac{\partial u_i}{\partial x_j}| \approx |\frac{\partial u_i}{\partial X_j}| << 1$ hold approximately. Neglecting the higher order terms in strain components, components of finite strain components in (2.11) and (2.12) are linearized and are reduced to

$$\varepsilon_{ij} = \frac{1}{2}\left(\frac{\partial u_i}{\partial x_j} + \frac{\partial u_j}{\partial x_i}\right) \tag{2.13}$$

or

$$\varepsilon_{xx} = \frac{\partial u_x}{\partial x}, \ \varepsilon_{yy} = \frac{\partial u_y}{\partial y}, \ \varepsilon_{zz} = \frac{\partial u_z}{\partial z},$$

$$\varepsilon_{yz} = \frac{1}{2}\left(\frac{\partial u_y}{\partial z} + \frac{\partial u_z}{\partial y}\right), \ \varepsilon_{zx} = \frac{1}{2}\left(\frac{\partial u_z}{\partial y} + \frac{\partial u_x}{\partial z}\right), \ \varepsilon_{xy} = \frac{1}{2}\left(\frac{\partial u_x}{\partial y} + \frac{\partial u_y}{\partial x}\right). \tag{2.14}$$

This is the assumption of small strains, and gives the bases for infinitesimal strain theory. The sign of a strain component is the same as that of a stress. The strain components ε_{11}, ε_{22}, ε_{33} with the same indices are normal strains and those ε_{ij} with different indices are shear strains. Because the strain is symmetric, $\varepsilon_{ij} = \varepsilon_{ji}$, and six components are independent. It is noted here, the shear strain components in (2.13) are of tensor strains and are a half of engineering shear strain components γ_{ij}. Tensor strain components are arranged in a matrix form as

$$\left(\varepsilon_{ij}\right) = \begin{bmatrix} \varepsilon_{11} & \varepsilon_{12} & \varepsilon_{13} \\ \varepsilon_{21} & \varepsilon_{22} & \varepsilon_{23} \\ \varepsilon_{31} & \varepsilon_{32} & \varepsilon_{33} \end{bmatrix} = \begin{bmatrix} \varepsilon_{11} & \frac{1}{2}\gamma_{12} & \frac{1}{2}\gamma_{13} \\ \frac{1}{2}\gamma_{21} & \varepsilon_{22} & \frac{1}{2}\gamma_{23} \\ \frac{1}{2}\gamma_{31} & \frac{1}{2}\gamma_{32} & \varepsilon_{33} \end{bmatrix}. \tag{2.15}$$

Equation (2.13) or (2.14) is the kinematic equations for infinitesimal strain theory providing the linear relationship between the displacement and the strain.

Exercise 2.3 Using the strain vector of independent six infinitesimal strain components defined as $\boldsymbol{\varepsilon} = \begin{bmatrix} \varepsilon_{xx} \ \varepsilon_{yy} \ \varepsilon_{zz} \ \gamma_{yz} \ \gamma_{zx} \ \gamma_{xy} \end{bmatrix}^T = \begin{bmatrix} \varepsilon_{xx} \ \varepsilon_{yy} \ \varepsilon_{zz} \ 2\varepsilon_{yz} \ 2\varepsilon_{zx} \ 2\varepsilon_{xy} \end{bmatrix}^T$, confirm that kinematic equations (2.14) is written as a matrix equation of $\boldsymbol{\varepsilon} = \boldsymbol{\partial}^T\mathbf{u}$, where $\boldsymbol{\partial}$ is that in Exercise 2.2. Strain vector uses engineering shear components γ_{ij} instead of tensor components ε_{ij}.

Fig. 2.4 Framework of force, displacement, stress and strain

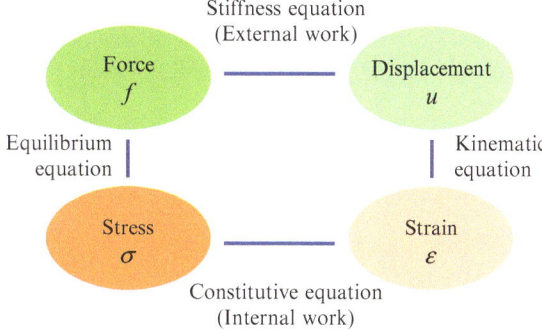

2.1.3   Constitutive Equations: Linear Elasticity

The stress components σ_{ij} are connected to force components by equilibrium equations as is in Sect. 2.1.1 and the strain components ε_{ij} are related to the displacement components through kinematic equations in Sect. 2.1.2. In order to explore the external work by force and displacement in a solid in terms of stress and strain, we need the third equations connecting stress and strain, that is, the constitutive equations. This equation gives the framework for the internal work, as is illustrated in Fig. 2.4.

In one-dimensional cases illustrated in Sect. 1.2.1, the stress is described as a linear function of strain for some cases reasonably, or as a nonlinear function of strain for other cases. Therefore, a general expression will be

$$\sigma_{ij} = \sigma_{ij}(\varepsilon_{kl}) \tag{2.16}$$

connecting the stress components σ_{ij} and the small strain components ε_{kl}. When the stress is linearly related to strain, the constitutive equation is reduced to

$$\sigma_{ij} = C_{ijkl}\varepsilon_{kl} \tag{2.17}$$

in tensor form or

$$\boldsymbol{\sigma} = \mathbf{C}\boldsymbol{\varepsilon} \tag{2.18}$$

in matrix form. This is the three-dimensional expression of Hooke's law (cf. (1.1) for one-dimensional case), and the coefficients C_{ijkl} and $\mathbf{C}$ are constants characterizing material properties. The solid body that obeys this constitutive equation is a linear elastic solid that is also called as Hookean. The number of independent constants of C_{ijkl} and $\mathbf{C}$ is twenty-one at most due to the symmetry of elastic coefficient, but is dependent on the properties of solid. The linear elastic constitutive equation is frequently used for hard tissues such as bone and teeth. It is also adopted for soft tissues as far as the deformation is small and the materials exhibit weak nonlinearity that can be approximated by a linear function.

2.1.3.1 Hooke's Law for Isotropy

For a linear elastic solid of isotropy having the same properties in all directions, only two coefficients are independent and the coefficients are written as

$$C_{ijkl} = \lambda \delta_{ij} \delta_{kl} + \mu (\delta_{ik} \delta_{jl} + \delta_{il} \delta_{jk}) \tag{2.19}$$

or

$$\mathbf{C} = \begin{bmatrix} \mathbf{C}_\sigma & \mathbf{0} \\ \mathbf{0} & \mathbf{C}_\tau \end{bmatrix}, \mathbf{C}_\sigma = \begin{bmatrix} \lambda + 2\mu & \lambda & \lambda \\ \lambda & \lambda + 2\mu & \lambda \\ \lambda & \lambda & \lambda + 2\mu \end{bmatrix},$$

$$\mathbf{C}_\tau = \begin{bmatrix} \mu & 0 & 0 \\ 0 & \mu & 0 \\ 0 & 0 & \mu \end{bmatrix}$$

in terms of Lame's constants λ and μ with Kronecher-δ tensor.[5] These constants are transformed from and to engineering elastic constants of Young's modulus (modulus of longitudinal elasticity) E, Poisson's ratio v, shear modulus (modulus of transverse elasticity) G and bulk modulus K as

$$\lambda = \frac{Ev}{(1+v)(1-2v)}, \mu = \frac{E}{2(1+v)}, \tag{2.21}$$

$$E = \frac{\mu(3\lambda + 2\mu)}{\lambda + \mu}, v = \frac{\lambda}{2(\lambda + \mu)}, G = \mu, K = \lambda + \frac{2}{3}\mu. \tag{2.22}$$

The isotropic constitutive equation (2.19) is then identical to a set of equations

$$\varepsilon_{xx} = \frac{1}{E}[\sigma_{xx} - v(\sigma_{yy} + \sigma_{zz})], \gamma_{xy} = 2\varepsilon_{xy} = \frac{1}{G}\sigma_{xy},$$

$$\varepsilon_{yy} = \frac{1}{E}[\sigma_{yy} - v(\sigma_{xx} + \sigma_{zz})], \gamma_{yz} = 2\varepsilon_{yz} = \frac{1}{G}\sigma_{yz},$$

$$\varepsilon_{zz} = \frac{1}{E}[\sigma_{zz} - v(\sigma_{xx} + \sigma_{yy})], \gamma_{zx} = 2\varepsilon_{zx} = \frac{1}{G}\sigma_{zx} \tag{2.23}$$

---

[5] Kronecker-δ tensor is defined as $\delta_{ij} = \begin{cases} 1 & \text{when} & i=j \\ 0 & \text{when} & i \neq j \end{cases}$. It corresponds unit matrix, e.g. $(\delta_{ij}) = \begin{bmatrix} 1 & 0 & 0 \\ 0 & 1 & 0 \\ 0 & 0 & 1 \end{bmatrix}$ in three dimensions, and replaces the index as $\delta_{ij}a_i = a_j$.

of a conventional strain–stress expression where the shear modulus is dependent on E and v as

$$G = \mu = \frac{E}{2(1+v)}. \tag{2.24}$$

Most biosolids have a hierarchical tissue structure and assumptions of homogeneity and isotropy are not satisfied precisely. The isotropy, however, provides the good basis for the first order analysis when the primal concern is for a macroscopic mechanical behavior.

2.1.3.2 Hooke's Law for Transverse Isotropy

A solid is transversely isotropic when its properties are the same in all directions within a plane but different in the direction perpendicular to the plane. In this case, the elastic moduli in the transverse plane are different from those in the perpendicular direction. For a linear elastic solid of the transverse isotropy, by choosing z-axis direction as the direction of anisotropy, the constitutive relations are written as

$$
\begin{aligned}
\varepsilon_{xx} &= \frac{1}{E}\left[\sigma_{xx} - v\sigma_{yy}\right] - \frac{v'}{E'}\sigma_{zz}, \gamma_{xy} = 2\varepsilon_{xy} = \frac{1}{G}\sigma_{xy}, \\
\varepsilon_{yy} &= \frac{1}{E}\left[\sigma_{yy} - v\sigma_{xx}\right] - \frac{v'}{E'}\sigma_{zz}, \gamma_{yz} = 2\varepsilon_{yz} = \frac{1}{G'}\sigma_{yz}, \\
\varepsilon_{zz} &= \frac{1}{E'}\sigma_{zz} - \frac{v'}{E}\left[\sigma_{xx} + \sigma_{yy}\right], \gamma_{zx} = 2\varepsilon_{zx} = \frac{1}{G'}\sigma_{zx}.
\end{aligned} \tag{2.25}
$$

where the shear modulus G is again dependent on E and v as (2.24). There are five independent constants. The elastic modulus E is the transverse Young's modulus $E_T = E = E_x = E_y$, and the modulus E' is the longitudinal Young's modulus $E_L = E' = E_z$. The transverse isotropy is widely used to characterize the property of one directionally reinforced solid of engineering materials. Some categories of biosolids have fiber structure along one direction and the transverse isotropy works as a reasonable model for them.

Exercise 2.4 Reorganize the constitutive relations (2.25) in coefficient tensor C_{ijkl} of (2.17) or coefficient matrix C of (2.18).

2.1.3.3 Hooke's Law for Orthotropy

For a solid of orthotropy or orthogonal anisotropy, the properties along each of three axes mutually orthogonal are defined. In this case, the constitutive relations are generalized as

$$\varepsilon_{xx} = \frac{1}{E_1}\sigma_{xx} - \frac{v_{21}}{E_2}\sigma_{yy} - \frac{v_{31}}{E_2}\sigma_{zz}, \gamma_{zx} = 2\varepsilon_{zx} = \frac{1}{G_{31}}\sigma_{zx},$$

$$\varepsilon_{yy} = \frac{1}{E_2}\sigma_{yy} - \frac{v_{12}}{E_1}\sigma_{xx} - \frac{v_{32}}{E_3}\sigma_{zz}, \gamma_{yz} = 2\varepsilon_{yz} = \frac{1}{G_{23}}\sigma_{yz},$$

$$\varepsilon_{zz} = \frac{1}{E_3}\sigma_{zz} - \frac{v_{13}}{E_1}\sigma_{xx} - \frac{v_{23}}{E_2}\sigma_{yy}, \gamma_{zx} = 2\varepsilon_{zx} = \frac{1}{G_{31}}\sigma_{zx} \qquad (2.26)$$

by using three Young's moduli, six Poisson's ratios and three shear moduli. It is noted that only three of six Poisson's ratios are independent because of three equations

$$\frac{v_{12}}{E_1} = \frac{v_{21}}{E_2}, \frac{v_{23}}{E_2} = \frac{v_{32}}{E_3}, \frac{v_{31}}{E_3} = \frac{v_{13}}{E_1}. \qquad (2.27)$$

Thus, the orthotropic Hooke's law has nine independent constants. The tubular structure is one of fundamental geometries of organs. In accordance with tubular structure, such organs frequently exhibit different characteristics in the longitudinal, radial and circumferential directions, established on a local coordinate system having mutually orthogonal axes. The orthotropy is, thus, a common framework for biomechanical analysis.

Exercise 2.5 Derive constraints among constants of orthotropy needed to satisfy the transversely isotropy.

2.1.4 Constitutive Equations: Nonlinear Elasticity

Nonlinearity is one of the fundamental characteristics of soft biosolids. The linearized simplification is not satisfactory even for the loads and deformations within a physiological range. Soft biosolids are viscoelastic in nature and exhibits hysteresis. Cyclic force/deformation such as in breathing, beating and walking is a common pattern in physiological external loading conditions, where a steady cyclic response is observed in relevant organs/tissues. A steady cyclic response is also reconstructed in experimental conditions after a several number of loading and unloading cycles that is called preconditioning. A nonlinear elasticity is thus considered for the loading process of a steady cyclic response, and another nonlinear elasticity is then considered for the unloading process. This is the concept of pseudo-lasticity, in which individual path is treated as nonlinear elasticity although the solid is not truly elastic.

2.1.4.1 Stress and Strain for Finite Deformation

For nonlinear behavior of finite deformation, analysis of deformation obeys the finite strain theory using Green strain tensor or Almansi strain tensor of (2.11) and (2.12). In accordance with the strain measure, different stress measures are used to consider the difference of areas in undeformed and deformed configurations, as mentioned in Sect. 1.2.1 for one dimension. For the internal force df_i acting on an area element dS of unit normal vector v_i, the force vector is related to the true stress components as

$$df_i = p_i dS, \; p_i = v_j \sigma_{ji} \tag{2.28}$$

here, the second equation is called Cauchy's formula. The internal force df_{0i} is similarly defined for the area element dS_0 having unit normal vector v_{0i} at the undeformed configuration. These two internal forces are related as

$$df_{0i} = df_i \tag{2.29}$$

similar to engineering-type traction and

$$df_{0i} = \frac{\partial X_i}{\partial x_j} df_j \tag{2.30}$$

to fictitious traction from a mathematical point of view. In accordance with these two internal force on area element dS_0, two stresses t_{ij} and s_{ij} are defined as

$$\frac{df_{0i}}{dS_0} = \frac{df_i}{dS_0} \equiv v_{0j} t_{ji} \tag{2.31}$$

and

$$\frac{df_{0i}}{dS_0} = \frac{\partial X_i}{\partial x_j} \frac{df_i}{dS_0} \equiv v_{0j} s_{ji}. \tag{2.32}$$

The former t_{ij} is the first Piola-Kirchhoff stress and the latter s_{ij} is the second Piola-Kirchhoff stress. These are related each other as

$$t_{ji} = \frac{\partial x_j}{\partial X_k} s_{ki} \; \text{ or } \; s_{ji} = \frac{\partial X_i}{\partial x_k} t_{jk} \tag{2.33}$$

and converted from and to Cauchy (true) stress as

$$t_{ji} = \frac{\rho_0}{\rho} \frac{\partial X_j}{\partial x_k} \sigma_{ki}, \; \sigma_{ji} = \frac{\rho}{\rho_0} \frac{\partial x_j}{\partial x_k} t_{ki} \tag{2.34}$$

and

$$s_{ji} = \frac{\rho_0}{\rho} \frac{\partial X_j}{\partial x_m} \frac{\partial X_i}{\partial x_n} \sigma_{mn}, \; \sigma_{ji} = \frac{\rho}{\rho_0} \frac{\partial x_j}{\partial X_m} \frac{\partial x_i}{\partial X_n} s_{mn} \tag{2.35}$$

respectively,[6] where ρ_0 and ρ are the densities at the undeformed and deformed configuration. The first Piola-Kirchhoff i.e. Lagrangian stress is not symmetric in general, while the second Piola-Kirchhoff stress is symmetric. The definition of s_{ji} is based on the fictitious traction of (2.32) and explaining physical meaning is difficult. However, any stress components are calculated from the second Piola-Kirchhoff stress s_{ji}, which mathematically describes the energy function with the Green strain E_{ji}. This is the reason why the second Piola-Kirchhoff stress is often used with the Green strain for the analysis of nonlinear elasticity. The equilibrium equation is then written as

$$\frac{\partial}{\partial X_i}\left(\frac{\partial x_j}{\partial X_m} s_{mi}\right) + \rho_0 G_{0j} = 0 \tag{2.36}$$

where G_{0j} is the body force per unit mass at the undeformed configuration converted from that at the deformed configuration G_j and $G_j = G_{0j}$ by conservation law.

Exercise 2.6 Examine the one-dimensional stresses in (1.10) as a uni-axial condition of (2.33) and (2.43).

2.1.4.2 Hyperelasticity

The constitutive equation for nonlinear elasticity of a finite strain is thus written as

$$s_{ij} = s_{ij}(E_{kl}) \tag{2.37}$$

between the second Piola-Kirchhoff stress tensor and Green strain tensor, although the choice of stress and strain pair of a constitutive equation is not restricted. This is a straightforward extension of constitutive equation of linear elasticity. Nevertheless, strain energy-based approach has been employed as an alternative. The strain energy density function $\rho_0 W$ is defined for a unit reference volume of biosolid as an analytical function of Green strain using the strain energy density W per unit mass.[7] Then the components of s_{ij} are calculated as derivatives of strain energy density $\rho_0 W$ with respect to corresponding strain components

$$s_{ij} = \frac{\partial(\rho_0 W)}{\partial E_{ij}} \tag{2.38}$$

---

[6] The ratio $\frac{\rho_0}{\rho}$ is reciprocal of the volume ratio $J = \frac{V}{V_0}$ of volume element deformed and undeformed because of conservation law $V_0 \rho_0 = V\rho$. cf. (2.42).

[7] Here the strain energy density function W is defined first for a unit mass, then that for a unit volume is represented by $\rho_0 W$. The strain energy function for a unit volume comes first is an alternative as employed in other textbook.

and this works as the constitutive equation connecting stress and strain. Solid body of constitutive relationship (2.38) is called a hyperelastic body. Here, the selection of nonlinear stress–strain relation in (2.37) is replaced with the selection of strain energy function. In selecting a strain energy density function, both of symmetric strain components E_{ij} and E_{ji} are separately involved so that $\rho_0 W$ has a symmetric form of strain components. The other stresses are

$$t_{ji} = \frac{\partial x_j}{\partial X_k} \frac{\partial(\rho_0 W)}{\partial E_{ki}} \tag{2.33}'$$

and

$$\sigma_{ji} = \frac{\rho}{\rho_0} \frac{\partial x_j}{\partial X_m} \frac{\partial x_i}{\partial X_n} \frac{\partial(\rho_0 W)}{\partial E_{mn}} \tag{2.35}'$$

Or, the first Piola-Kirchhoff, i.e. Lagrange stress is directly derived from the strain energy density function as

$$t_{ji} = \frac{\partial(\rho_0 W)}{\partial(\partial x_j / \partial X_i)}. \tag{2.39}$$

Strain components are dependent on the choice of coordinates but the invariants are not. Hence, the strain density function is written as a function of invariants instead of components of strain, for an isotropic body. The Green strain tensor E_{ij} defined by (2.11) is written by using the deformation gradient tensor $F_{ij} = \frac{\partial x_i}{\partial X_j}$ or the right Cauchy–Green deformation tensor

$$C_{ij} = \frac{\partial x_k}{\partial X_i} \frac{\partial x_k}{\partial X_j} = F_{ki} F_{kj}$$

as

$$E_{ij} = \frac{1}{2}(F_{ki} F_{kj} - \delta_{ij}) = \frac{1}{2}(C_{ij} - \delta_{ij}). \tag{2.40}$$

The strain energy density function of isotropic hyper-elasticity is mainly described in terms of principal invariants of the right Cauchy–Green deformation tensor C_{ij}

$$I_1 = C_{ii} = 3 + 2E_{ii}$$

$$I_2 = \frac{1}{2}(C_{ii}C_{jj} - C_{ij}C_{ij}) = 3 + 4E_{ii} + 2(E_{ii}E_{jj} - E_{ij}E_{ji})$$

$$I_3 = \det[C_{ij}] = \det[\delta_{ij} + 2E_{ij}]. \tag{2.41}$$

The incompressibility is described by using the determinant of deformation gradient F_{ij} or the strain invariant I_3 as

$$J = \det[F_{ij}] = 1 \quad \text{or} \quad I_3 = (\det[F_{ij}])^2 = 1. \tag{2.42}$$

For further details, refer textbooks for mechanics of nonlinear elasticity (e.g. Holzapfel 2000; Taber 2004).

Exercise 2.7 When the strain energy density function is a function of strain invariants I_1, I_2 and I_3, the constitutive equation is written as

$$s_{ij} = \frac{\partial(\rho_0 W)}{\partial E_{ij}} = \frac{\partial(\rho_0 W)}{\partial I_1}\frac{\partial I_1}{\partial E_{ij}} + \frac{\partial(\rho_0 W)}{\partial I_2}\frac{\partial I_2}{\partial E_{ij}} + \frac{\partial(\rho_0 W)}{\partial I_3}\frac{\partial I_3}{\partial E_{ij}} \tag{2.43}$$

in terms of derivatives of invariants. Using the chain rule for these derivatives e.g. $\frac{\partial I_1}{\partial E_{ij}} = \frac{\partial I_1}{\partial C_{kl}}\frac{\partial C_{kl}}{\partial E_{ij}}$ and the relation

$$\frac{\partial C_{kl}}{\partial E_{ij}} = \frac{\partial(2E_{kl} + \delta_{kl})}{\partial E_{ij}} = 2\delta_{ki}\delta_{lj}, \tag{2.44}$$

show the following derivatives

$$\frac{\partial I_1}{\partial C_{ij}} = \delta_{ij}, \frac{\partial I_2}{\partial C_{ij}} = I_1\delta_{ij} - C_{ij} \text{ and } \frac{\partial I_3}{\partial C_{ij}} = I_2\delta_{ij} - I_1 C_{ij} + C_{ik}C_{kj} = I_3 C_{ij}^{-1}, \tag{2.45}$$

and derive $\frac{\partial I_1}{\partial E_{ij}}, \frac{\partial I_2}{\partial E_{ij}}$ and $\frac{\partial I_3}{\partial E_{ij}}$ for (2.43).

2.2 Mechanical Properties of Bone

Mechanical properties are the most essential element for the biomechanical modeling for computational analysis, although much attention has been paid for the geometrical modeling. The section provides the choice of constitutive model equation and representative model constants. The constitutive model is not determined uniquely even for a specific tissue, and its constants are very dependent on the sample specimen examined and the individual. Therefore, it should be noticed that specific values for mechanical properties shown in the below are not fundamental physical constants in pure science, but representative biomechanical engineering constants.

2.2.1 Cortical Bone

For hard tissue of bone and teeth, the major concern falls in the linear elastic range of mechanical behavior, except for the cases related to impact response or failure. The bone tissue is generally classified into cortical (compact) bone and cancellous

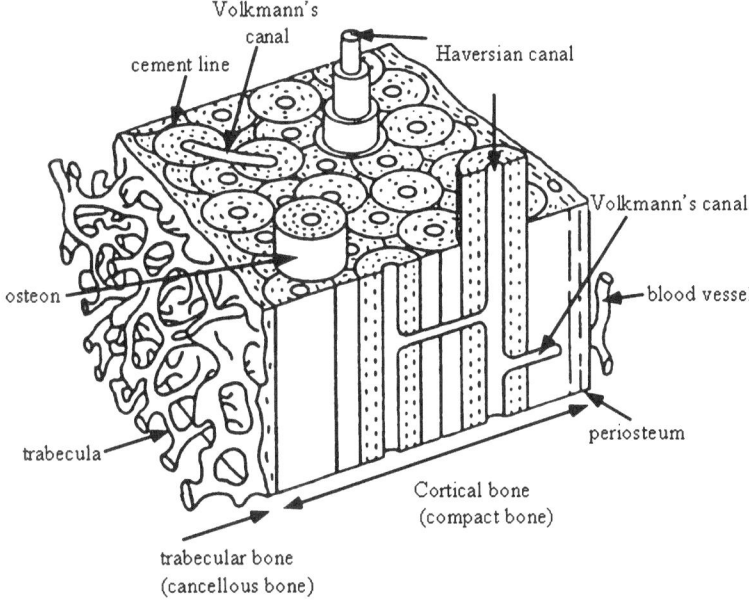

Fig. 2.5   Schematic illustration of cortical bone

(trabecular or spongy) bone. The cortical bone is compact and macroscopically a continuum body. Thus the linear elasticity works as the constitutive model of the cortical bone. The simplest model is the isotropic linear elastic body as is mentioned in Sect. 2.1.3, and the material constants of isotropic linear elastic solid is two engineering constants, Young's modulus and Poisson's ratio.

Cortical bone in diaphysis of long bones has anisotropic tissue structure as schematically illustrated in Fig. 2.5, and the mechanical properties are also not isotropic. Thus, the transverse isotropy or orthotropy is the candidate of the constitutive model of cortical bone. The transverse isotropy assumes the isotropy in the plane of radial and circumferential directions and the longitudinal direction is the anisotropic one by referring to its microstructure. A set of material constants is

$$E_1 = 12.0,\ E_2 = 13.4,\ E_3 = 20.0\,(\text{GPa})$$
$$G_{12} = 4.53,\ G_{13} = 5.61,\ G_{23} = 6.23\,(\text{GPa})$$
$$\nu_{12} = 0.376,\ \nu_{13} = 0.222,\ \nu_{23} = 0.235$$
$$\nu_{21} = 0.422,\ \nu_{31} = 0.371,\ \nu_{32} = 0.350 \tag{2.46}$$

for human femoral bone by ultrasound technique (Ashman et al. 1984), where directions 1, 2 and 3 indicate the radial, circumferential and longitudinal directions respectively. Other set will be found in Buskirk et al. (1981). For human tibia, a set of constants is

$$E_1 = 6.91,\ E_2 = 8.51,\ E_3 = 18.4\,(\text{GPa})$$
$$G_{12} = 2.42,\ G_{13} = 3.56,\ G_{23} = 4.91(\text{GPa})$$
$$v_{12} = 0.49,\ v_{13} = 0.12,\ v_{23} = 0.14$$
$$v_{21} = 0.62,\ v_{31} = 0.32,\ v_{32} = 0.31 \tag{2.47}$$

obtained by mechanical testing (Knets and Malmeisters 1977 or Cowin 1983). A different set of constants for human tibia is given with standard deviations as

$$E_1 = 7.0 \pm 0.2,\ E_2 = 8.7 \pm 0.3,\ E_3 = 18.7 \pm 0.4\,(\text{GPa})$$
$$G_{12} = 2.46 \pm 0.14,\ G_{13} = 3.63 \pm 0.18,\ G_{23} = 5.01 \pm 0.17\,(\text{GPa})$$
$$v_{12} = 0.49 \pm 0.02,\ v_{13} = 0.12 \pm 0.01,\ v_{23} = 0.14 \pm 0.01$$
$$v_{21} = 0.62 \pm 0.02,\ v_{31} = 0.31 \pm 0.01,\ v_{32} = 0.31 \pm 0.01 \tag{2.48}$$

by combining ultrasound technique and mechanical testing (Knets 2002).

These data sets support the orthotropic constitutive equation as the model of cortical bone. However, transverse isotropic constitutive equation is a reasonable alternative for the mechanical analysis, because the difference between the radial and circumferential directions is small compared to that from the longitudinal direction. Experimental data of material constants assuming the transverse isotropy are also available as

$$E_1 = E_2 = 11.5,\ E_3 = 17.0\,(\text{GPa})$$
$$G_{13} = G_{23} = 3.28\,(\text{GPa})$$
$$v_{12} = v_{21} = 0.58,\ v_{31} = v_{32} = 0.46 \tag{2.49}$$

where standard deviations are reported for several subsets of specimens mechanically tested (Reilly and Burstein 1975).

The values of material constants as the anisotropic elastic body scatter widely experiment by experiment, specimen by specimen and individual by individual. Therefore, the isotropic linear elasticity is frequently assumed for the cortical bone in the computational biomechanics analysis despite that the cortical bone has anisotropic properties (e.g. Cilingir et al. 2007; Laville et al. 2009).

These material properties are, of course, dependent on the material composition. In order to take this aspect into account, the elastic modulus is related to the bone density by a linear regression analysis as

$$\text{Longitudinal modulus } E = 14261 \times density - 13430\ (\text{MPa})$$
$$\text{Transverse modulus } E = 4979 \times density - 3122\ (\text{MPa}) \tag{2.50}$$

for human proximal femur where the density is given in g/cm^3 and the moduli in MPa (Lotz et al. 1991). Medical image indices are also used to characterize the elastic properties. For instance, the mean Young's modulus is correlated to the volumic bone mineral density (*vBMD*) as

$$E = 0.025vBMD - 4.58 \, \text{GPa}$$
$$vBMD = 0.428\rho_{HA} + 243.23. \tag{2.51}$$

by linear regression where *vBMD* is measured by DXA and given in mg/cm^3 and the equivalent X-ray CT density ρ_{HA} is given in mg/ml HA (Duchemin et al. 2008). These provide us a method for personalized modeling of mechanical properties in the form of density–modulus relationship by regression with linear and power functions for image-based computational analysis (e.g. Austman et al. 2008).

It is noted here that cortical bone does not behaves as an elastic body in the strict sense but does as a viscoelastic body. Nevertheless the elastic body assumption is acceptable within the range of strain rate in daily life activity[8] and widely used for the computational analyses and simulations.

Second point to be noted is that any constitutive model based on data fitting by regression analysis, back (or inverse) analysis and others has its own confidential interval. The interpolation is generally acceptable, but the exterpolation outside the confidential interval needs careful consideration.[9]

Exercise 2.8 Calculate the elastic constants such as G_{12} not given explicitly in (2.49) assuming the transverse isotropy.

2.2.2   Cancellous Bone

Cancellous bone has a three-dimensional lattice or network structure made of trabeculae, that is, column, beam or plate like elements, and is much porous than cortical bone (Fig. 2.6). Thus the mechanical properties of cancellouse bone is obviously different from that of cortical bone, and is known to be influenced by the degree of porosity (e.g. Carter and Hayes 1976; Turner and Cowin 1987). One of the most well-known reports on the elastic modulus of cancellous bone is

$$E = 3790\dot{\varepsilon}^{0.06}\rho^3 \, (\text{MPa}) \tag{2.52}$$

---

[8] The stress–strain curve is strain-rate dependent and is approximated by Ramberg–Osgood model. That is, Young's modulus is described as a power function of strain rate, but the effect of strain rate on it is as much as 15% in daily activity (Cowin 1989).

[9] The concept of a confidential interval in data fitted to the constitutive model and the problem of exterpolation are not limited to the constitutive models of cortical bone in this section. These points become more severely important for constitutive models for soft tissues exhibiting a strong nonlinearity found in the following sections.

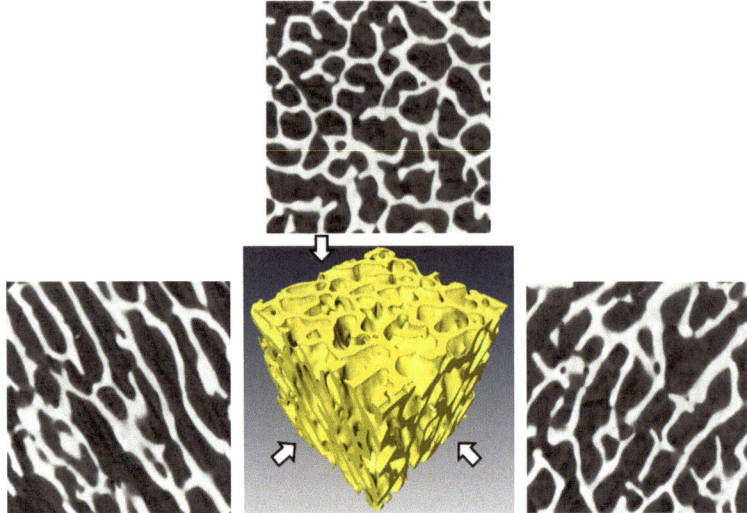

Fig. 2.6 Cancellous bone. Three-dimensional CT image

where ρ is the apparent density in kg/m^3 and $\dot{\varepsilon}$ is the strain rate in the inverse of second (Carter and Hayes 1977). This is a power law regression between elastic modulus and apparent bone density. The expression of

$$E = A\rho^B \text{ or } \log E = \log A + B \log \rho \qquad (2.53)$$

has then been a standard model as well as the linear regression model. This power law model was extensively studied for specific exponents of $B = 2$ and/or 3 (Rice et al. 1988), for the mixture supported by some theoretical considerations of a cellular structure (Gibson 1985) and for other exponents. For an assumed physiological strain rate, the model of (2.52) is reduced to the density–modulus relationship of (2.53) with a constant coefficient of

$$A = 2875 \text{ for } \dot{\varepsilon} = 0.01 \text{ and } A = 3071 \text{ for } \dot{\varepsilon} = 0.03 \qquad (2.54)$$

and exponent $B = 3$. Further studies showed the exponent is closer to 2 than 3 (Hodgskinson and Currey 1992; Yang et al. 1999). Morgan et al. (2003) showed several pairs of constants for different anatomical sites for the elastic modulus E in MPa and the apparent density ρ in g/cm^3 as follows:

$$A = 4730, B = 1.56 \text{ for vertebra}$$
$$A = 15520, B = 1.93 \text{ for proximal tibia}$$
$$A = 15010, B = 2.18 \text{ for greater trochanter}$$
$$A = 6850, B = 1.49 \text{ for femoral neck}$$
$$A = 8920, B = 1.83 \text{ for pooled} \qquad (2.55)$$

It should be noted that each pair of constants has its range of apparent density. There are many data on the linear regression between the elastic modulus and the apparent density, but the power regression produces statistically better fits to many data (e.g. Ciarelli et al. 1991; Rho et al. 1995). The power function model of (2.53) was modified as

$$E = AV_f{}^B \alpha^C \tag{2.56}$$

in which introduced are the bone volume fraction V_f=bone volume/bulk volume and the ash fraction α=ash mass/dry bone mass with constants

$$A = 84.37, \, B = 2.58 \pm 0.02, \, \text{and} \, C = 2.74 \pm 0.13 \tag{2.57}$$

by Hernandez et al. (2001) Medical imaging index is also adopted to the power function model such as

$$E = 1310\rho^{1.4} \, \text{and} \, E = 0.5\rho_{K_3HPO_4}^{1.2} \tag{2.58}$$

with modulus E in MPa, apparent density ρ in g/cm^3 and equivalent mineral density $\rho_{K_3HPO_4}$ in mg/cm^3 K$_3$HPO$_4$ (Lotz et al. 1990).

This kind of regression model by a power function or a linear function for elastic modulus–density relation is basically for the isotropic linear elastic constitutive model. The trabecular architecture of cancellous bone has a lattice-like structure with characteristic orientations of trabeculae, and this is an aspect to be known as the Wolff's law (Wolff 1892, 1986, 2010). Despite that this fact explicitly implies the anisotropy of the mechanical property of cancellous bone as a macroscopic continuum body, the isotropic linear elasticity has been widely accepted as a constitutive model of cancellous bone for computational analyses. The medical imaging technology such as micro CT in these days is able to reconstruct the trabecular architecture as it is, and the modulus of trabecula becomes important for computational analyses as well as the modulus of cancellous bone (e.g. Turner et al. 1999; Zysset et al. 1999). These informations connect the mechanical properties of cancellous bone as a macroscopic continuum from/to those of trabecula solid as a microscopic continuum through micro finite element model and analysis (e.g. Muller and Puegsegger 1996; van Rietbergen et al. 1998).

Comprehensive model and data on bone will be found in several books (e.g. Cowin 1989, 2001; Martin et al. 1998; Currey 2002).

Exercise 2.9 Draw the density–modulus curves by the power-law in (2.53) with constants given by (2.55), and compare them.

2.3   Material Properties of Soft Tissue

The material properties of soft tissue are significantly dependent on the type of organ and tissue. This section picks up a couple of soft tissues and describes strain energy functions used to provide constitutive equation in the context of hyperelasticity.

These are not strain energy but pseudo strain energy for pseudo-hyperelasticity in a strict sense because of the inelasticity of soft tissue material in general. Nevertheless the use of pseudo strain energy is useful to analyze and understand the soft tissue behaviour quantitatively. In this section, the term "strain energy" is sometimes used instead of the pseudo strain energy for convenience.

2.3.1 Arterial Wall

Arterial wall (Fig. 2.7) has a cylindrical shape and the radial, circumferential and longitudinal directions are selected for the axes of a local coordinate system where the local Cartesian coordinate system is defined along these directions. The choice of function for pseudo strain energy is not unique and there are many varieties. The guideline for the selection is the smaller number of parameters and a better fit to experimental data from an engineering point of view for computational analyses.

Two different types of pseudo strain energy density functions are examined by Fung et al. (1979). The first specific form is

$$\rho_0 W = AE_{\theta\theta}^2 + BE_{\theta\theta}E_{zz} + CE_{zz}^2 + DE_{\theta\theta}^3 + EE_{\theta\theta}^2 E_{zz} + FE_{\theta\theta}E_{zz}^2$$
$$+ GE_{zz}^3 \tag{2.59}$$

of a polynomial expression (Vaishnav et al. 1973) of Green strain components in two dimensions of circumferential (θ) and longitudinal (z) coordinates.[10] The second form is an exponential function

$$\rho_0 W = \frac{C}{2} \exp\left[a_1(E_{\theta\theta}^2 - E_{\theta\theta}^{*2}) + a_2(E_{zz}^2 - E_{zz}^{*2}) + 2a_4(E_{\theta\theta}E_{zz} - E_{\theta\theta}^*E_{zz}^*)\right] \tag{2.60a}$$

$$= \frac{C'}{2} \exp\left[a_1 E_{\theta\theta}^2 + a_2 E_{zz}^2 + 2a_4 E_{\theta\theta}E_{zz}\right] \tag{2.60b}$$

where the terms $()^*$ are reference strain components corresponding to an arbitrary selected physiological stress state. Constants reported include

$$A = \begin{cases} -14.7220 \\ -12.0062 \end{cases}, B = \begin{cases} -4.1606 \\ 5.1405 \end{cases}, C = \begin{cases} 4.4821 \\ -1.5936 \end{cases}, D = \begin{cases} 16.5753 \\ -2.1292 \end{cases},$$

$$E = \begin{cases} 29.9390 \\ 23.5706 \end{cases}, F = \begin{cases} 6.2090 \\ -7.3431 \end{cases}, G = \begin{cases} -1.8999 & \text{for lower aorta} \\ 2.2069 & \text{for upper aorta} \end{cases} \tag{2.61}$$

---

[10] Throughout Sect. 2.3, notations in the original articles are used as much as possible in order to make further reference to the details easier.

in $10 \, \text{kPa} = 10^5 \, \text{dyn/cm}^2$, and

$$C' = \begin{cases} 2.1744 \\ 3.3856 \end{cases} (10\text{kPa}), \ a_1 = \begin{cases} 9.5660 \\ 2.8173 \end{cases}, \ a_2 = \begin{cases} 3.0913 \\ 0.6239 \end{cases}, \ a_4 = \begin{cases} 0.8805 \\ -.5790 \end{cases}$$

with

$$E_{\theta\theta}^* = \begin{cases} 0.2743 \\ 0.4061 \end{cases}, \ E_{zz}^* = \begin{cases} 0.6495 \\ 0.9566 \end{cases}, \ S_{\theta\theta}^* = \begin{cases} 4.9 \\ 5.2 \end{cases} (10\text{kPa}),$$

$$S_{zz}^* = \begin{cases} 3.0 \\ 1.9 \end{cases} (10\text{kPa}) \quad \begin{array}{l} \text{for lower aorta} \\ \text{for upper aorta} \end{array} \tag{2.62}$$

both for leporine lower aorta in the first raw of (2.61) and (2.62) and for upper aorta in the second raw. The smaller variation of constants in the exponential function may suggest preference of the model of (2.60a). Exponential type strain energy function in three dimensions is also used as is found in Wang et al. (2006).

A strain energy function of decoupled representation is proposed by Holzapfel and Weizacker (1998). The energy has two terms as

$$\rho_0 W = \Psi = \Psi_{iso} + \Psi_{aniso} \tag{2.63}$$

with

$$\Psi_{iso} = c_1(I_1 - 3), \ \Psi_{aniso} = c_2 \exp[Q - 1] \tag{2.64a, b}$$

and

$$Q = a_1 E_1^2 + a_2 E_2^2 + 2a_4 E_1 E_2. \tag{2.64c}^{[11]}$$

Indices 1 and 2 denote the circumferential and longitudinal directions respectively. The first term of strain energy is for the isotropic part and is represented by the neo-Hookean body. The second term is for the anisotropic part and represented by an exponential type with exponent of the second order polynomial. Here the strain in the third direction is coped with the incompressibility condition. The presence of the isotropic part is the essential difference from the previous model of (2.60b). The constants are determined as

---

[11] Green strain components E_1 and E_2 denotes the normal strain components $E_{11} = E_{\theta\theta}$ in circumferential direction x_1 and $E_{22} = E_{zz}$ in longitudinal direction x_2, respectively.

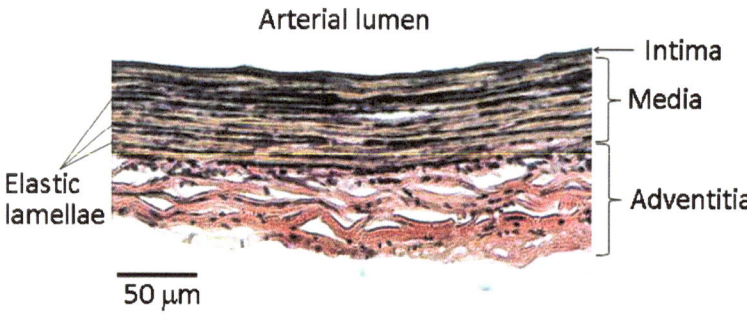

Fig. 2.7 Cross-section of arterial wall (leporine carotid artery)

$$c_1 = \begin{cases} 30523.0 \\ 0 \end{cases} (\text{Pa}), \; c_2 = \begin{cases} 430.791 \\ 30582.00 \end{cases} (\text{Pa})$$

$$a_1 = \begin{cases} 5.36603 \\ 0.6823782 \end{cases}, \; a_2 = \begin{cases} 3.55858 \\ 0.6055760 \end{cases},$$

$$a_4 = \begin{cases} -0.317206 & \text{for two-term s.e.} \\ 0.6116318 & \text{for only-second-term s.e.} \end{cases} \tag{2.65}$$

for the two-term strain energy function (the first raw) and the only-second-term strain energy (the second raw) for rat abdominal aorta, and the advantage of two-term strain energy function is more appreciable for wide range of blood pressure. More sophisticated hyperelastic modeling of an arterial wall considering the micro-structure is found in the review by Gasser et al. (2005). Wang et al. (2006) illustrated the two-layer model. A study for various strain energy functions including advanced ones is found in Holzapfel et al. (2000) where constitutive models for arterial wall are compared.

Computational procedure such as a finite element analysis inevitably requests in the curve fitting process in order to determine unknown parameters of strain energy function. Therefore, a direct analysis with an assumed constitutive relationship and the back (inverse) analysis for identification of a constitutive relationship should be harmoniously combined with the computational biomechanics analysis.

Exercise 2.10 Show the full set of constants of an exponential-type pseudo strain energy function of (2.60b) from the set of constants given in (2.62).

2.3.2 Skin

Skin is a membranous organ and its behaviour is modeled by two-dimensional constitutive relation for plane stress condition. This is not an isotropic body due to the presence of Langer's line to which the collagen and elastic fibers are preferentially aligned. We will find a list of pseudo elastic model for the stretch in Lanir (1987). One of pseudo strain energy functions of orthotropy is

$$\rho_0 W = \frac{1}{2}\{\alpha_1 E_1{}^2 + \alpha_2 E_2{}^2 + 2\alpha_4 E_1 E_2\}$$
$$+ \frac{c}{2}\exp\left[a_1 E_1{}^2 + a_2 E_2{}^2 + 2a_4 E_1 E_2 + \gamma_1 E_1{}^3 + \gamma_2 E_2{}^3 + \gamma_4 E_1{}^2 E_2 + \gamma_5 E_1 E_2{}^2\right]$$

(2.66)

given by Tong and Fung (1976). Constants for this strain energy function are shown for a case of (a) $\alpha_1 = \alpha_2$ and $\gamma_1 = \gamma_2 = \gamma_4 = \gamma_5 = 0$ and another case of (b) $\alpha_1 = \alpha_2$, $\gamma_1 = \gamma_2 = 0$ and $\gamma_4 = \gamma_5$ as

$$a_1 = \left\{\begin{array}{l} 3.79 \\ 3.79 \end{array}\right., a_2 = \left\{\begin{array}{l} 12.7 \\ 18.4 \end{array}\right., a_4 = \left\{\begin{array}{l} 0.587 \\ 0.587 \end{array}\right., c = \left\{\begin{array}{l} 0.00794 \\ 0.00794 \end{array}\right.,$$
$$\alpha_1 = \alpha_2 = \left\{\begin{array}{l} 10.4 \\ 10.4 \end{array}\right., S_{\alpha_4}^* = \left\{\begin{array}{l} 2.59 \\ 2.59 \end{array}\right., \gamma_4 = \gamma_5 \left\{\begin{array}{ll} 0 & \text{for case a)} \\ 15.6 & \text{for case b)} \end{array}\right.$$

(2.67)

for a sample specimen of leporine abdominal skin. It was demonstrated that these two sets of constants successfully reproduced the experimental stretch-force curve and the cubic terms in the exponent has no significance in it.

A modified strain energy function

$$\rho_0 W = \alpha_1 E_1{}^2 + \alpha_2 E_2{}^2 + \alpha_3 E_1 E_2 + c\exp\left[A_1 E_1{}^2 + A_2 E_2{}^2\right] - p(J - 1) \quad (2.68)$$

was used for the mechanical response of *in vivo* human skin (Flynn et al. 2011). In this energy function, any cross term of strain components are neglected in the exponent and the incompressibility condition is added as the last term, where J is the volume ratio (cf. (2.42) and footnote 6) and p is the Lagrange multiplier corresponding to the hydrostatic pressure. The constants determined for anterior forearm skin and posterior upper arm skins *in vivo* include

$$\alpha_1 = \left\{\begin{array}{l} 15.961 \\ 1.563 \end{array}\right.(\text{kPa}), \alpha_2 = \left\{\begin{array}{l} 0.048 \\ 8772 \end{array}\right.(\text{kPa}), \alpha_3 = \left\{\begin{array}{l} 38.574 \\ 12.355 \end{array}\right.(\text{kPa})$$
$$c = \left\{\begin{array}{l} 0.0117 \\ 0.1000 \end{array}\right.(\text{Pa}), A_1 = \left\{\begin{array}{l} 31.403 \\ 51.089 \end{array}\right., A_2 = \left\{\begin{array}{ll} 24.400 & \text{for forearm} \\ 7.962 & \text{for upper arm} \end{array}\right.$$

(2.69)

reproducing the reaction curve for different directions.

2.3.3 Cornea

Cornea is a collagenous tissue and its extrafibrillar matrix is primarily water. It behaves highly nonlinearly and viscoelastically. The hyperelasty is a candidate for constitutive model in the sense of pseudo elasticity. Among hyperelastic modeling, Elsheikh et al. (2006) employed Ogden's strain energy function

$$\rho_0 W = U = \sum_{i=1}^{N} \frac{2\mu_i}{\alpha_i^2} \left(\lambda_1^{\alpha_i} + \lambda_2^{\alpha_i} + \lambda_3^{\alpha_i} - 3 \right) \tag{2.70}$$

with the incompressibility condition $J - 1 = 0$ by means of the Lagrange multiplier of an ocular model. They assumed the order of the model as $N = 4$ and determined the constants as

$$\mu_1 = -110.3, \ \mu_2 = -55.64, \ \mu_3 = 108.2, \ \mu_4 = -53.54,$$
$$\alpha_1 = 14.97, \alpha_2 = 16.06, \ \alpha_3 = 12.93, \ \alpha_4 = 11.99 \tag{2.71}$$

to reproduce stress–strain curve data with initial modulus of $E = 0.30$Pa (Elsheikh and Anderson 2005). Ogden's model is for isotropic body, and the strain energy function of a generalized Mooney–Rivlin solid of

$$\rho_0 W = \sum_{i+j=1}^{N} c_{ij}(I_1 - 3)^i (I_2 - 3)^j \tag{2.72}$$

is an another strain energy model for isotropic body. Roy and Dupps (2011) used a strain energy function with two coefficients c_{10} and c_{20}, where the anisotropy of cornea is taken by choosing these coefficients as a function of meridians.

Decoupled representation of strain energy (cf. (2.63) for arterial wall) is also available for the cornea tissue. Decoupling is for the volumetric (dilational) deformation and the isochoric (distortional) deformation as

$$\rho_0 W = \Psi = \Psi_{vol} + \Psi_{isoch} \tag{2.73}$$

in the context of quasi-incompressible hyperelasticity (Alastrue et al. 2006). The volumetric part of strain energy was given as

$$\Psi_{vol} = \frac{1}{D} [\ln (J)]^2 \tag{2.74}$$

where D^{-1} is the penalty coefficient for $\ln(J) = 0$ i.e. $J = 1$. The isotropic contribution to an isochoric part of the strain energy was then the first order terms of (2.72) and the anisotropic contribution was expressed as

$$\Psi_{isoch} = \Psi_{isoch}^{matrix} + \Psi_{isoch}^{fibrils} \tag{2.75a}$$

$$\Psi_{isoch}^{matrix} = \frac{C_1}{2} (\hat{I}_1 - 3) + \frac{C_2}{2} (\hat{I}_2 - 3) \tag{2.75b}$$

$$\Psi_{isoch}^{fibrils} = \sum_{l=4,6} \frac{k_1}{2k_2} \exp[k_2(\hat{I}_l - 1)^2 - 1] \tag{2.75c}$$

where the isotropic part represents tissue matrix and irregularly arranged fibers such as in Bowman's layer and the anisotropic part does the lattice-like arrangement of fibers in corneal stroma. The specific function of (2.75c) for $\Psi^{fibrils}_{isoch}$ comes from a model of an arterial wall (Holzapfel et al. 2000), and each term is written as a function of modified invariant $\hat{I}_4 = a_i \hat{C}_{ij} a_j$ or $\hat{I}_6 = b_i \hat{C}_{ij} b_j$ with unit vector (a_i) or (b_i) which represents fiber orientation.[12] The constants for human cornea used by Alastrue et al. (2006) are

$$D = 13.3333 \, \text{MPa}^{-1}, \; C_1 = 0.005 \, \text{MPa} \; , \; C_2 = 0.0 \, \text{MPa},$$

$$k_1 = 0.004852 \, \text{MPa and } k_2 = 102.643. \tag{2.76}$$

Functions Ψ_{vol}, Ψ_{iso} and Ψ_{aniso} are not limited to those shown here and have varieties. Especially, anisotropic part have been extended to consider the distribution of fiber orientation (Pandolfi and Holzapfel 2008; Studer et al. 2009).

2.4   Principles of Virtual Work and Stationary Potential Energy

The principles of virtual work and stationary potential energy are variational form of boundary value problem for equilibrium of solid body. These provide bases of computational analysis and simulation of mechanical behavior of deformable body by the finite element method that is most commonly used method in commercially available package. These principles are described for the infinitesimal strain elasticity and for finite strain elasticity.

2.4.1   Boundary Value Problem for Equilibrium

The equilibrium equation for infinitesimal (small) strain elasticity is written in the index notation and in a matrix notation as

$$\sigma_{ji,j} + b_i = 0 \text{ and } \boldsymbol{\partial\sigma} + \mathbf{b} = \mathbf{0} \text{ in } \Omega \tag{2.77}^{13}$$

---

[12] This constitutive model uses the modified right Cauchy–Green tensor $\hat{C}_{ij}$, and $\hat{I}_l$ denotes their invariants. Some fundamental variables in modified components are given by $F_{ij} = J^{1/3}\hat{F}_{ij}$, $C_{ij} = J^{2/3}\hat{C}_{ij}$, $\hat{C}_{ij} = \hat{F}_{ki}\hat{F}_{kj}$, $E_{ij} = J^{2/3}E_{ij} + \frac{1}{2}(J^{2/3} - 1)\delta_{ij}$ and $\hat{E}_{ij} = \frac{1}{2}(\hat{C}_{ij} - \delta_{ij})$.

[13] The notation $(\,)_{,i}$ stands for the partial derivative $\frac{\partial()}{\partial x_i}$ with respect to the coordinates x_i. That is, $\sigma_{ji,j} = \frac{\partial(\sigma_{ji})}{\partial x_j}$ with summation convention.

using the stress vector $\boldsymbol{\sigma}$ in the Voigt notation and the differential operator matrix $\boldsymbol{\partial}$ (cf. Exercise 2.2) for the matrix notation. This differential equation accompanies the traction boundary condition for stress

$$t_i = \sigma_{ji}n_j = t_i^* \text{ and } \mathbf{t} = \mathbf{v}^T\boldsymbol{\sigma} = \mathbf{t}^* \text{ on } \Gamma_t \tag{2.78}$$

and the displacement boundary condition

$$u_i = u_i^* \text{ and } \mathbf{u} = \mathbf{u}^* \text{ on } \Gamma_u \tag{2.79}$$

where $()^*$ indicates the specified value, n_j stands for the unit normal vector and the surface domain $\Gamma = \Gamma_t + \Gamma_u$ denotes the boundary of the volumetric domain Ω of the solid body. The set of three equations constitutes a boundary value problem.

The constitutive equation (2.16) relates stress $\boldsymbol{\sigma}$ and strain $\boldsymbol{\varepsilon}$ such as

$$\sigma_{ij} = \sigma_{ij}(\varepsilon_{kl}) \text{ and } \boldsymbol{\sigma} = \boldsymbol{\sigma}(\varepsilon) \tag{2.80}$$

and the kinematic equation (2.13) relates the strain $\boldsymbol{\varepsilon}$ to the displacement $\mathbf{u}$ as

$$\varepsilon_{kl} = \tfrac{1}{2}(u_{k,l} + u_{l,k}) \text{ and } \boldsymbol{\varepsilon} = \boldsymbol{\partial}^T\mathbf{u} \tag{2.81}$$

(see Exercise 2.3). Therefore the displacement u is the unknown variable of the boundary value problem.

Exercise 2.11 Show the 3 by 6 matrix $\mathbf{v}^T$ of (2.78).

Exercise 2.12 Derive the differential equation for displacement u_i from (2.77) and (2.81) with the constitutive equation (2.19) of Hooke's law for isotropic linear elastic body. This is known as the Navier (or Cauchy-Navier) equation of elasticity.

2.4.2 Principle of Virtual Work

The external virtual work done by the body force b_i and by the surface traction t_i^* with the virtual displacement δu_i is written by

$$\begin{aligned} \delta W_{external}(\mathbf{u}, \delta\mathbf{u}) &= \int_\Omega \delta u_i b_i d\Omega + \int_{\Gamma_t} \delta u_i t_i^* d\Gamma \\ &= \int_\Omega \delta\mathbf{u}^T \mathbf{b} d\Omega + \int_{\Gamma_t} \delta\mathbf{u}^T \mathbf{t}^* d\Gamma \end{aligned} \tag{2.82}$$

and the internal virtual work done by the stress σ_{ij} (tensor component) or σ_i (component of stress vector $\boldsymbol{\sigma}$) is done by

$$\delta W_{internal}(\mathbf{u}, \delta\mathbf{u}) = \int_\Omega \delta\varepsilon_{ij}\sigma_{ij} d\Omega = \int_\Omega \delta\boldsymbol{\varepsilon}^T \boldsymbol{\sigma} d\Omega \tag{2.83}$$

where $\delta\varepsilon_{ij} = \frac{1}{2}(\delta u_{i,j} + \delta u_{j,i})$ is virtual strain due to virtual displacement δu_i, and $\delta\boldsymbol{\varepsilon}$ is its vector notation. The principle of virtual work is now written as

$$\delta W_{internal}(\mathbf{u}, \delta\mathbf{u}) - \delta W_{external}(\mathbf{u}, \delta\mathbf{u}) = 0. \tag{2.84}$$

This is a variational form[14] of the boundary value problem and will satisfy (2.77) and (2.78).

In the case of a finite strain described in Sect. 2.1.4, the internal virtual work is written by using the second Piola-Kirchhoff stress tensor and the Green strain tensor as

$$\delta W_{internal}(u_i, \delta u_i) = \int_{\Omega_0} \delta E_{ij} s_{ij} d\Omega \tag{2.85}$$

and the external virtual work is by

$$\delta W_{external}(u_i, \delta u_i) = \int_{\Omega_0} \rho_0 G_{0i} \delta u_i d\Omega + \int_{\Gamma_{0t}} P^*_{0i} \delta u_i d\Gamma. \tag{2.86}$$

The principle of virtual work is

$$\int_{\Omega_0} \delta E_{ij} s_{ij} d\Omega = \int_{\Omega_0} \rho_0 G_{0i} \delta u_i d\Omega + \int_{\Gamma_{0t}} P^*_{0i} \delta u_i d\Gamma \tag{2.87}$$

It is noted here that the body domain for integral of the internal work is of the undeformed configuration for a finite strain problem but not deformed one, although these two are not distinguished for the infinitesimal (small) strain problem. This note is applicable for the volumetric and surface domains of external work as a functional of virtual displacement. The definitions of body force, surface traction and surface normal vector also refer to the undeformed configuration and is explicitly distinguished by the suffix $()_0$[15]

Exercise 2.13 Show that the principle of virtual work (2.87) for finite strain is written as

$$\int_{\Omega_0} \left\{ \frac{\partial}{\partial X_i} \left(\frac{\partial x_j}{\partial X_m} s_{mi} \right) + \rho_0 G_{0j} \right\} \delta u_j d\Omega + \int_{\Gamma_{0t}} \left(P^*_{0i} - \frac{\partial x_j}{\partial X_m} s_{mi} n_{0i} \right) \delta u_i d\Gamma = 0. \tag{2.88}$$

---

[14] This is also called as a weak form.

[15] The body force G_{0i} per unit mass is explained at (2.36). The surface traction P_{0i} is such that $P_{0i} dS_0 = \frac{\partial x_j}{\partial X_m} s_{mi} n_{0i} dS_0$ for the boundary Γ_{0t} where n_{0i} is the surface unit normal to the undeformed configuration.

Use the divergence theorem (Gauss's theorem) with

$$u_j + \delta u_j = u_j^* \text{ i.e. } \delta u_j = 0 \text{ on } \Gamma_{0u} = \Gamma_0 - \Gamma_{0t}C \tag{2.89}$$

and

$$\frac{\partial x_j}{\partial X_m} s_{mi} \frac{\partial \delta u_j}{\partial X_i} = \left(\delta_{jm} + \frac{\partial u_j}{\partial X_m} \right) s_{mi} \frac{\partial \delta u_j}{\partial X_i}$$

$$= \frac{1}{2} \left\{ \left(\delta_{jm} + \frac{\partial u_j}{\partial X_m} \right) s_{mi} \frac{\partial \delta u_j}{\partial X_i} + \left(\delta_{jm} + \frac{\partial u_j}{\partial X_m} \right) s_{im} \frac{\partial \delta u_j}{\partial X_i} \right\} = \delta E_{mi} s_{mi} \tag{2.90}$$

with the symmetry of $s_{mi} = s_{im}$.

2.4.3   Principle of Stationary Potential Energy

In the case of linear elasticity, the strain energy density function is a quadratic form of strain as

$$U(\varepsilon_{ij}) = \frac{1}{2} \varepsilon_{ij} \sigma_{kl} = \frac{1}{2} \varepsilon_{ij} C_{ijkl} \varepsilon_{kl}$$

$$U(\boldsymbol{\varepsilon}) = \frac{1}{2} \boldsymbol{\varepsilon}^T \boldsymbol{\sigma} = \frac{1}{2} \boldsymbol{\varepsilon}^T \mathbf{C} \boldsymbol{\varepsilon} \tag{2.91}$$

from which the constitutive equation (2.17) or (2.18) is derived as

$$\sigma_{ij} = \frac{\partial U(\varepsilon_{ij})}{\partial \varepsilon_{ij}} \text{ or } \boldsymbol{\sigma} = \frac{\partial U(\boldsymbol{\varepsilon})}{\partial \boldsymbol{\varepsilon}} \tag{2.92}$$

The total potential energy Π is

$$\Pi = \Pi_{internal} + \Pi_{external} \tag{2.93}$$

$$\Pi_{internal} = \int_\Omega U(\varepsilon_{ij}) d\Omega \tag{2.94a}$$

$$\Pi_{external} = - \int_\Omega u_i b_i d\Omega - \int_{\Gamma_t} u_i t_i^* d\Gamma$$

$$= - \int_\Omega \mathbf{u}^T \mathbf{b} d\Omega - \int_{\Gamma_t} \mathbf{u}^T \mathbf{t}^* d\Gamma \tag{2.94b}$$

and is a functional of displacement u_i since the strain is a variable dependent on a displacement. The principle of virtual work is derived as the stationary condition of total potential energy with respect to displacement:

$$\text{stationary } \Pi(u_i) \text{ with respect to } u_i \text{ subject to (2.79)} \tag{2.95a}$$

i.e.

$$\delta\Pi(u_i) = 0 \tag{2.95b}$$

This principle of the stationary potential energy (2.95a) works for a hyperelastic body as well. By using the potential energies, it is given by

$$\Pi_{internal} = \int_{\Omega_0} \rho_0 W d\Omega \tag{2.96a}$$

$$\Pi_{external} = -\int_{\Omega_0} \rho_o b_i u_i d\Omega - \int_{\Gamma_{0t}} P_{0i}^* u_i d\Gamma \tag{2.96b}$$

in a similar manner for the principle of virtual work for a finite deformation for the case of conservative (dead) loads. The principle of virtual work (2.87) is immediately derived from $\delta\Pi = 0$, i.e. the stationary condition of Π. For further details, textbooks for variational principles by Oden and Reddy (1976), Washizu (1982) and others are helpful.

Exercise 2.14 Derive the principle of virtual work in (2.87) from the principle of stationary potential energy equation (2.95b) with (2.96a,b) for finite strain hyperelasticity.

2.5 Finite Element Method

2.5.1 Finite Element Discretization and Approximation

The domain of solid body Ω is discretized into small subdomains Ω_e named finite elements so that

$$\Omega = \bigcup_e \Omega_e \quad \phi = \bigcap_e \Omega_e \tag{2.97}$$

The displacement field $u_i(\mathbf{x})$ in the entire domain Ω is approximated by the set of a displacement field $u_i^{(e)}(\mathbf{x})$ locally defined for each finite element Ω_e. A set of points $\mathbf{x}_N \ N = 1, ..., N_e$ identifies a finite element, and these points are called nodes. For an element-wise displacement field $u_i(\mathbf{x})$ for element e is then approximately represented as

$$u_i(\mathbf{x}) = \Phi_N^{(e)}(\mathbf{x})u_{iN}^{(e)} \text{ in } \Omega_e,$$

or

$$\mathbf{u}(\mathbf{x}) = \boldsymbol{\Phi}^{(e)}(\mathbf{x})\mathbf{u}^{(e)} \text{ in } \Omega_e \qquad (2.98)^{16}$$

by means of a convenient function interpolating the displacement at the nodes $u_{iN}^{(e)} = u_i(\mathbf{x}_N)$ where $\mathbf{u}^{(e)} = \left[\mathbf{u}_1^{(e)\mathrm{T}} ... \mathbf{u}_N^{(e)\mathrm{T}} ... \mathbf{u}_{Ne}^{(e)\mathrm{T}}\right]^{\mathrm{T}}$ is the nodal displacement vector of the element e. The interpolation function $\Phi_N^{(e)}$ is named shape function, and satisfies

$$\Phi_N^{(e)}(\mathbf{x}_M) = \delta_{NM} \text{ and } \sum_N \Phi_N^{(e)}(\mathbf{x}) = 1 \qquad (2.99)$$

so that the interpolated displacement coincide with the nodal displacement itself at each node as $u_{iM}^{(e)} = u_i^{(e)}(\mathbf{x}_M) = \Phi_N^{(e)}(\mathbf{x}_M)u_{iN}^{(e)}$ and so that the constant displacement in an element is represented as is.

2.5.2 Finite Element Equation for Small Strain Linear Elasticity

The strain tensor ε_{ij} in a finite element e becomes

$$\varepsilon_{ij} = \frac{1}{2}\left(\frac{\partial \Phi_N^{(e)}(\mathbf{x})}{\partial x_j}u_{iN}^{(e)} + \frac{\partial \Phi_M^{(e)}(\mathbf{x})}{\partial x_i}u_{jM}^{(e)}\right) \qquad (2.100a)$$

in component expression and

$$\boldsymbol{\varepsilon} = \mathbf{B}^{(e)}(\mathbf{x})\mathbf{u}^{(e)}, \quad \mathbf{B}^{(e)} = \partial \boldsymbol{\Phi}^{(e)} \qquad (2.100b)$$

in a matrix expression in accordance with the interpolation of displacement. Substituting the interpolation of displacement (2.98) and that of strain (2.100a) into the energy of (2.94a,b), the energies for a finite element e written as

$$\Pi_{internal}^{(e)} = \frac{1}{2}k_{iNkM}^{(e)}u_{iN}^{(e)}u_{kM}^{(e)}, \qquad (2.101a)$$

---

[16] Summation convention is applied for the nodal index N but not for the element index (e) in parentheses. That is, (2.98) is identical to $u_i^{(e)}(\mathbf{x}) = \sum_{N=1}^{Ne} \Phi_N^{(e)}(\mathbf{x})u_{iN}^{(e)}$ in conventional expression.

$$\Pi_{external}^{(e)} = -f_{iN} u_{iN}^{(e)} \tag{2.101b}$$

by using element stiffness components $k_{iNkM}^{(e)}$

$$k_{iNkM}^{(e)} = \int_{\Omega_e} C_{ijkl} \frac{\partial \Phi_N^{(e)}}{\partial x_j} \frac{\partial \Phi_M^{(e)}}{\partial x_i} \tag{2.102}$$

and equivalent nodal force components f_{iN}

$$f_{iN} = \int_{\Omega_e} \Phi_N^{(e)} b_i d\Omega + \int_{\Gamma_{te}} \Phi_N^{(e)} t_i^* d\Gamma \tag{2.103}$$

where the domains for integration are limited to those of the element. This is also written in a matrix form as

$$\Pi_{internal}^{(e)} = \frac{1}{2} \mathbf{u}^{(e)T} \mathbf{k}^{(e)} \mathbf{u}^{(e)}, \ \Pi_{external}^{(e)} = -\mathbf{f}^{(e)T} \mathbf{u}^{(e)} \tag{2.104}$$

by using the element stiffness (symmetric) matrix $\mathbf{k^{(e)}}$ and the nodal force vector $\mathbf{f}^{(e)}$ equivalent to the body force and surface traction

$$\mathbf{k}^{(e)} = \int_{\Omega_e} \mathbf{B^{(e)T}} \mathbf{C} \mathbf{B}^{(e)} d\Omega, \ \mathbf{f}^{(e)} = \int_{\Omega_e} \boldsymbol{\Phi}^{(e)T} \mathbf{b} d\Omega + \int_{\Gamma_{te}} \boldsymbol{\Phi}^{(e)T} \mathbf{t}^* d\Gamma. \tag{2.105}$$

The total potential energy $\Pi^{(e)}$ for an element is a quadratic form of nodal displacements $u_{iN}^{(e)}$ or $\mathbf{u}^{(e)}$ and thus the stationary condition $\delta \Pi^{(e)} = 0$ is reduced to a set of linear equations as

$$\frac{\partial \Pi^{(e)}}{\partial u_{iN}^{(e)}} = k_{iNkM}^{(e)} u_{kM}^{(e)} - f_{kM}^{(e)} = 0 \ \text{ or } \ \frac{\partial \Pi^{(e)}}{\partial \mathbf{u}^{(e)}} = \mathbf{k}^{(e)} \mathbf{u}^{(e)} - \mathbf{f}^{(e)} = 0. \tag{2.106}$$

The total potential energy for the entire domain of Ω is the sum of $\Pi^{(e)}$

$$\Pi_{internal} = \sum_e \Pi_{internal}^{(e)} = \sum_e \frac{1}{2} k_{iNkM}^{(e)} u_{iN}^{(e)} u_{kM}^{(e)} \tag{2.107}$$

and

$$\Pi_{external} = \sum_e \Pi_{external}^{(e)} = -\sum_e \frac{1}{2} f_{iN} u_{iN}^{(e)}$$

is reorganized as

$$\Pi_{internal} = \frac{1}{2} K_{iPkQ} u_{iP} u_{kQ} \tag{2.108}$$

and

$$\Pi_{external} = -\frac{1}{2} F_{iP} u_{iP}$$

by paying attention to the fact that an element-wise index number N and M $N, M = 1, ..., N_e$ for (2.106) corresponds to the specific global index number $P, Q = 1, ..., N_t$ for (2.108) where N_t denotes the total number of nodes for entire domain Ω. This process is more explicitly written in matrix form as

$$\Pi = \sum_e \left(\frac{1}{2} \mathbf{u}^{(e)T} \mathbf{k}^{(e)} \mathbf{u}^{(e)} - \mathbf{u}^{(e)T} \mathbf{f}^{(e)} \right) = \frac{1}{2} \mathbf{U}^T \mathbf{K} \mathbf{U} - \mathbf{U}^T \mathbf{F}. \tag{2.109}$$

which relates the nodal displacement vector $\mathbf{u}^{(e)}$ of an element to the global nodal displacement vector $\mathbf{U} = \left[\mathbf{u}_1^T .... \mathbf{u}_{Nt}^T \right]^T$ as

$$\mathbf{u}^{(e)} = \mathbf{A}^{(e)} \mathbf{U}. \tag{2.110}$$

where a matrix $\mathbf{A}^{(e)}$ is defined for each finite element and its matrix elements are zero or one. The element stiffness matrices $\mathbf{k}^{(e)}$ and nodal force vectors $\mathbf{f}^{(e)}$ are assembled into the global stiffness matrix $\mathbf{K}$ and the global nodal force vector $\mathbf{F}$ as follows:

$$\mathbf{K} = \sum_e \left(\mathbf{A}^{(e)T} \mathbf{k}^{(e)} \mathbf{A}^{(e)} \right), \quad \mathbf{F} = \sum_e \left(\mathbf{A}^{(e)T} \mathbf{f}^{(e)} \right). \tag{2.111}$$

The resultant potential energy is originally a functional of unknown displacement field $u_i(\mathbf{x})$, but now is a function of nodal displacement u_{iP}. By means of a finite element approximation by element-wise interpolation, the stationary condition of total potential energy is

$$K_{iPkQ} u_{kQ} - F_{iP} = 0$$

$$\mathbf{K} \mathbf{U} - \mathbf{F} = 0. \tag{2.112}$$

This is a system of linear equations of unknown nodal displacements, and called as a system finite element equation and/or a global stiffness equation. The coefficient K_{iPkQ} is the system (global) stiffness and $\mathbf{K}$ is the global stiffness matrix. By taking the displacement boundary condition of (2.79) into account, the equations for the displacement u_{iP} of the node P on Γ_u are replaced with

$$u_{iP} = u_i^*(x_P) \quad \text{or} \quad \mathbf{u}_P = \mathbf{u}^*(\mathbf{x}_P). \tag{2.113}$$

The finite element equation (2.106) for an element and (2.112) for the entire system are also derived based on the principle of virtual work. The finite element interpolation (2.98) for displacement $\mathbf{u}$ is again used for the virtual displacement $\delta\mathbf{u}$ as

$$\delta\mathbf{u}(x) = \boldsymbol{\Phi}^{(e)}(\mathbf{x})\delta\mathbf{u}^{(e)} \text{ in } \Omega_e \tag{2.114}$$

and the virtual strain induced by virtual displacements is then

$$\delta\boldsymbol{\varepsilon}(\mathbf{x}) = \partial\boldsymbol{\Phi}^{(e)}(\mathbf{x})\delta\mathbf{u}^{(e)} = \mathbf{B}^{(e)}\delta\mathbf{u}^{(e)} \text{ in } \Omega_e \tag{2.115}$$

The internal and external virtual works for the element e are

$$\delta W_{internal}^{(e)}(\mathbf{u}, \delta\mathbf{u}) = \int_{\Omega_e} \delta\boldsymbol{\varepsilon}^T \boldsymbol{\sigma} d\Omega$$

$$= \int_{\Omega_e} \delta\mathbf{u}^{(e)T}\mathbf{B}^{(e)T}\mathbf{C}\mathbf{B}^{(e)}\mathbf{u}^{(e)} d\Omega = \delta\mathbf{u}^{(e)T}\mathbf{k}^{(e)}\mathbf{u}^{(e)} \tag{2.116}$$

and

$$\delta W_{external}^{(e)}(\mathbf{u}, \delta\mathbf{u}) = \int_{\Omega_e} \delta\mathbf{u}^T \mathbf{b} d\Omega + \int_{\Gamma_{te}} \delta\mathbf{u}^T \mathbf{t}^* d\Gamma$$

$$= \int_{\Omega_e} \delta\mathbf{u}^{(e)T}\boldsymbol{\Phi}^{(e)T}\mathbf{b} d\Omega + \int_{\Gamma_{te}} \delta\mathbf{u}^{(e)T}\boldsymbol{\Phi}^{(e)T}\mathbf{t}^* d\Gamma = \delta\mathbf{u}^{(e)T}\mathbf{f}^{(e)}. \tag{2.117}$$

The principle of virtual work for an element gives us

$$\delta\mathbf{u}^{(e)T}\mathbf{k}^{(e)}\mathbf{u}^{(e)} = \delta\mathbf{u}^{(e)T}\mathbf{f}^{(e)} \text{ i.e. } \mathbf{k}^{(e)}\mathbf{u}^{(e)} = \mathbf{f}^{(e)}. \tag{2.118}$$

The second equation is identical to (2.106). The same consideration for the principle of virtual work for the entire system gives us (2.112).

Once the nodal displacements $\mathbf{U}$ i.e. u_{iP} are obtained by solving the global finite element equation (2.112), the nodal displacements $\mathbf{u}^{(e)}$ i.e. $u_{iN}^{(e)}$ for each element e are extracted from $\mathbf{U}$ by (2.110) and the strain is calculated by (2.100b) and stress is then gained from the constitutive equation (2.18) for each finite element e. This is written as

$$\boldsymbol{\varepsilon} = \mathbf{B}^{(e)}(\mathbf{x})\mathbf{u}^{(e)} = \mathbf{B}^{(e)}(\mathbf{x})\mathbf{A}^{(e)}\mathbf{U} \text{ for element } e \tag{2.119}$$

$$\boldsymbol{\sigma} = \mathbf{C}\boldsymbol{\varepsilon} = \mathbf{C}\mathbf{B}^{(e)}(\mathbf{x})\mathbf{A}^{(e)}\mathbf{U} \text{ for element } e \tag{2.120}$$

in a matrix form. These are the stages of post-processing.

2.5.3   Finite Element Equation for Finite Strain Hyperelasticity

In accordance with the interpolation of (2.98), the deformation gradient, right Cauchy–Green deformation and Green strain tensors are written as

$$F_{ij} = \frac{\partial x_i}{\partial X_j} = \frac{\partial (X_i + u_i)}{\partial X_j} = \delta_{ij} + \frac{\partial \Phi_N^{(e)}}{\partial X_j} u_{iN}^{(e)}, \tag{2.121}$$

$$C_{ij} = F_{ki} F_{kj} = \left(\delta_{ki} + \frac{\partial \Phi_N^{(e)}}{\partial X_i} u_{kN}^{(e)} \right) \left(\delta_{kj} + \frac{\partial \Phi_M^{(e)}}{\partial X_j} u_{kM}^{(e)} \right) \tag{2.122}$$

and

$$E_{ij} = \frac{1}{2} \left(\frac{\partial \Phi_N^{(e)}}{\partial X_j} u_{iN}^{(e)} + \frac{\partial \Phi_N^{(e)}}{\partial X_i} u_{jN}^{(e)} + \frac{\partial \Phi_N^{(e)}}{\partial X_i} u_{kN}^{(e)} \frac{\partial \Phi_M^{(e)}}{\partial X_j} u_{kM}^{(e)} \right) \tag{2.123}$$

for a finite element e. Similarly the virtual displacement gradient is written by

$$\frac{\partial \delta u_i}{\partial X_j} = \frac{\partial \Phi_N^{(e)}}{\partial X_j} \delta u_{iN}^{(e)} \tag{2.124}$$

for the virtual displacement

$$\delta u_i(\mathbf{x}) = \Phi_N^{(e)}(\mathbf{x}) \delta u_{iN}^{(e)} \tag{2.125}$$

interpolated by using the same shape functions as those of displacements in (2.98).
   Substitution of (2.121) and (2.124) into (2.90) yields the internal virtual work written as

$$\begin{aligned} \delta W_{internal}^{(e)} &= \int_{\Omega_0} \delta E_{mj} s_{mj} d\Omega \\ &= \int_{\Omega_0} \left(\delta_{jm} + \frac{\partial \Phi_M^{(e)}}{\partial X_m} u_{jM}^{(e)} \right) s_{mi} \left(\frac{\partial \Phi_N^{(e)}}{\partial X_i} \delta u_{jN}^{(e)} \right) d\Omega. \end{aligned} \tag{2.126}$$

The external virtual work is then written as

$$\delta W_{external}^{(e)} = \int_{\Omega_0} \rho_0 G_{0j} \left(\Phi_N^{(e)}(\mathbf{x}) \delta u_{jN}^{(e)} \right) d\Omega + \int_{\Gamma_{0t}} P_{0i}^* \left(\Phi_N^{(e)}(\mathbf{x}) \delta u_{iN}^{(e)} \right) d\Gamma. \tag{2.127}$$

Removing the nodal virtual displacement $\delta u_{jN}^{(e)}$ from integrants, the principle of virtual work becomes

$$\int_{\Omega_0}\left(\delta_{jm}+\frac{\partial\Phi_{\mathrm{M}}^{(e)}}{\partial X_m}u_{jM}^{(e)}\right)s_{mi}\frac{\partial\Phi_{\mathrm{N}}^{(e)}}{\partial X_i}\mathrm{d}\Omega\delta u_{jN}^{(e)}$$

$$-\int_{\Omega_0}\rho_0 G_{0j}\Phi_{N}^{(e)}(\mathbf{x})\mathrm{d}\Omega\delta u_{jN}^{(e)}-\int_{\Gamma_{0t}}P_{0j}^*\Phi_{N}^{(e)}(\mathbf{x})\mathrm{d}\Gamma\delta u_{jN}^{(e)}=0. \qquad (2.128)$$

and yields the set of equations

$$\int_{\Omega_0}\left(\delta_{jm}+\frac{\partial\Phi_{\mathrm{M}}^{(e)}}{\partial X_m}u_{jM}^{(e)}\right)s_{mi}\frac{\partial\Phi_{\mathrm{N}}^{(e)}}{\partial X_i}\mathrm{d}\Omega$$

$$-\int_{\Omega_0}\rho_0 G_{0j}\Phi_{N}^{(e)}(\mathbf{x})\mathrm{d}\Omega-\int_{\Gamma_{0t}}P_{0j}^*\Phi_{N}^{(e)}(\mathbf{x})\mathrm{d}\Gamma=0. \qquad (2.129)$$

for a finite element e. This is a set of finite element equations for an element.

The stress s_{ij} is calculated from the strain energy density function $\rho_0 W$ as in (2.38) for a hyperelastic body. The strain energy density is a function of the Green strain E_{ij}, right Cauchy–Green deformation C_{ij} and/or deformation gradient F_{ij} tensors which are functions of nodal displacements u_{iN}, $N = 1, ..., N_e$. Therefore, the finite element equation (2.129) is a set of nonlinear equations of unknown nodal displacements u_{iN} of the element. The virtual works of all finite elements are summed up in a similar fashion as those from (2.104) to (2.108), resulting in the finite element equations for the system as the set of nonlinear equations of unknown nodal displacements at all the nodes, i.e. u_{iP}, $P = 1, ..., N_t$.

2.5.4 Shape Functions: Simplex, Complex and Multiplex Elements

The choice of a shape function is the fundamental stage for the finite element analysis in practice. This section explains the background for construction of shape function for simplex, complex and multiplex elements.

2.5.4.1 Simplex Elements

The simplex model is the simplest finite element, and is a very good element to understand the concept of a finite element approximation. For k-dimensional space, the simplex element has $N_e = k + 1$ nodes for interpolation (Fig. 2.8). The element-wise interpolation is of (2.98)

$$u_j(\mathbf{x})=\Phi_{N}^{(e)}(\mathbf{x})u_{jN}^{(e)} \text{ in } \Omega_e \qquad (2.98)$$

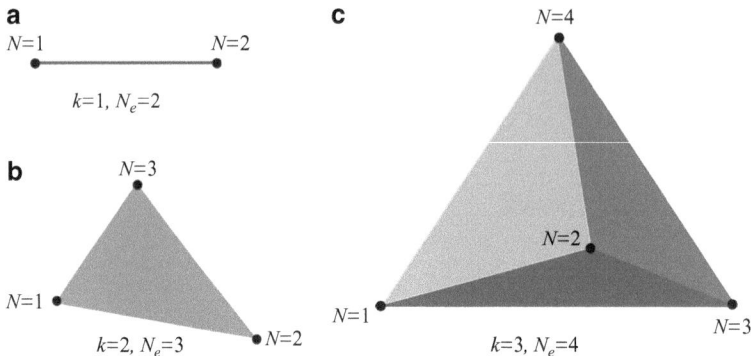

Fig. 2.8 Simplex elements in (**a**) one-, (**b**) two- and (**c**) three-dimensional space

where the index j of the degree of freedom is take from $\{1,...,k\}$ and the index N of nodes is from $\{1,..., k+1\}$ for the simplex element. The displacement field is thus the k-dimensional linear function

$$u_j(\mathbf{x}) = a_0 + a_i x_i. \tag{2.130}$$

This is the displacement function of a simplex model. The condition of the displacement at the nodes provides a set of linear equation for the coefficients a_i as

$$u_j(\mathbf{x}_N) = a_0 + a_i x_{iN} = u_{jN}. \tag{2.131}$$

For element e, the $k+1$ coefficients a_κ $\kappa = 0, ..., k$ are determined as a function of nodal coordinates $\mathbf{x}_N^{(e)} = (x_{iN}^{(e)})$ and nodal displacement u_{jN} by solving (2.131) for each j. The displacement u_j in the element is then reorganized in the form of (2.98). Using the notation of matrix calculus, (2.131) in a matrix form

$$\begin{pmatrix} 1 & x_{11}^{(e)} & x_{21}^{(e)} & \cdots & x_{k1}^{(e)} \\ 1 & x_{12}^{(e)} & x_{22}^{(e)} & \cdots & x_{k2}^{(e)} \\ \vdots & & & & \vdots \\ 1 & x_{1Ne}^{(e)} & x_{2Nw}^{(e)} & \cdots & x_{kNe}^{(e)} \end{pmatrix} \begin{pmatrix} a_0 \\ a_1 \\ \vdots \\ a_k \end{pmatrix} = \begin{pmatrix} u_{j1} \\ u_{j2} \\ \vdots \\ u_{jNe} \end{pmatrix} \tag{2.132}$$

directly gives

$$\Phi_N^{(e)} = \alpha_N + \beta_{iN} x_i \tag{2.133}$$

with

$$\alpha_N = \frac{\text{cofactor}(D_{N1})}{\det \mathbf{D}}, \ \beta_{iN} = \frac{\text{cofactor}(D_{N(i+1)})}{\det \mathbf{D}} \tag{2.134}$$

where cofactor(D_{Ni})denotes the D_{Ni} cofactor (adjunct) of the coefficient matrix $\mathbf{D}$ of (2.132). We have the same shape functions $\Phi_N^{(e)}$ regardless of the displacement component u_j because of the same nodal coordinates $x_{iN}^{(e)}$ for all components u_j.

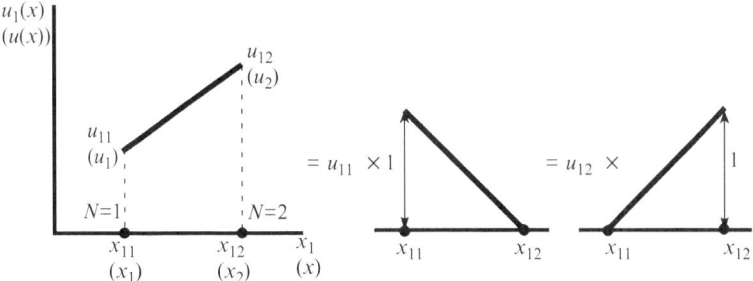

Fig. 2.9 Finite element approximation by one-dimensional simplex element

The simplex element of one-dimension is a line segment with two end nodes as shown in Fig. 2.8a. It has one coordinate variable x_1 and one displacement component u_1, and the displacement function is

$$u_1(\mathbf{x}) = a_0 + a_1 x_1. \tag{2.135}$$

Using the nodal coordinates $x_{11}^{(e)}$ of node $N = 1$ and $x_{12}^{(e)}$ of node $N = 2$ of an element e, (2.131) becomes

$$\begin{pmatrix} 1 & x_{11}^{(e)} \\ 1 & x_{12}^{(e)} \end{pmatrix} \begin{pmatrix} a_0 \\ a_1 \end{pmatrix} = \begin{pmatrix} u_{11}^{(e)} \\ u_{12}^{(e)} \end{pmatrix}. \tag{2.136}$$

in the matrix form. The shape functions

$$\Phi_1^{(e)} = \frac{x_{12}^{(e)} - x}{x_{12}^{(e)} - x_{11}^{(e)}} \text{ and } \Phi_2^{(e)} = \frac{x - x_{11}^{(e)}}{x_{12}^{(e)} - x_{11}^{(e)}} \tag{2.137}$$

are linear functions of the Lagrange interpolation.[17] The interpolation by one-dimensional simplex element is illustrated in Fig. 2.9.

For two-dimensional simplex element (Fig. 2.8b), $k = 2$, $N_e = 3$ and the element is of triangle shape. The displacement function for u_j $j = 1, 2$ is

---

[17] In non-index notation for coordinate system, we have only the nodal index for u and x. Displacement function is $u(x) = a_0 + a_1 x$, and (2.132) is $\begin{pmatrix} 1 & x_1 \\ 1 & x_2 \end{pmatrix} \begin{pmatrix} a_0 \\ a_1 \end{pmatrix} = \begin{pmatrix} u_1 \\ u_2 \end{pmatrix}$. Thus $\begin{pmatrix} a_0 \\ a_1 \end{pmatrix} = \frac{1}{x_2 - x_1} \begin{pmatrix} x_2 & -x_1 \\ -1 & 1 \end{pmatrix} \begin{pmatrix} u_1 \\ u_2 \end{pmatrix}$ and shape functions are $\Phi_1^{(e)} = \frac{x_2 - x}{x_2 - x_1}$ and $\Phi_2^{(e)} = \frac{x - x_1}{x_2 - x_1}$. These gives us the finite element approximation $u(x) = \frac{x_2 - x}{x_2 - x_1} u_1 + \frac{x - x_1}{x_2 - x_1} u_2$ or $u(x) = \begin{pmatrix} \frac{x_2 - x}{x_2 - x_1} & \frac{x - x_1}{x_2 - x_1} \end{pmatrix} \begin{pmatrix} u_1 \\ u_2 \end{pmatrix}$.

$$u_j(\mathbf{x}) = a_0 + a_1 x_1 + a_2 x_2 \tag{2.138}$$

Equation (2.132) for two-dimension

$$\begin{pmatrix} 1 & x_{11}^{(e)} & x_{21}^{(e)} \\ 1 & x_{12}^{(e)} & x_{22}^{(e)} \\ 1 & x_{13}^{(e)} & x_{23}^{(e)} \end{pmatrix} \begin{pmatrix} a_0 \\ a_1 \\ a_2 \end{pmatrix} = \begin{pmatrix} u_{j1} \\ u_{j2} \\ u_{j3} \end{pmatrix}. \tag{2.139}$$

results the shape functions as

$$\Phi_N^{(e)} = \alpha_N + \beta_{iN} x_i \tag{2.140}$$

with

$$(\alpha_N) = \frac{1}{2A} \begin{pmatrix} x_{12}^{(e)} x_{23}^{(e)} - x_{13}^{(e)} x_{22}^{(e)} \\ x_{13}^{(e)} x_{21}^{(e)} - x_{11}^{(e)} x_{23}^{(e)} \\ x_{11}^{(e)} x_{22}^{(e)} - x_{12}^{(e)} x_{21}^{(e)} \end{pmatrix},$$

$$(\beta_{iN}) = \frac{1}{2A} \begin{pmatrix} x_{22}^{(e)} - x_{23}^{(e)} & x_{23}^{(e)} - x_{21}^{(e)} & x_{21}^{(e)} - x_{22}^{(e)} \\ x_{13}^{(e)} - x_{12}^{(e)} & x_{11}^{(e)} - x_{13}^{(e)} & x_{12}^{(e)} - x_{11}^{(e)} \end{pmatrix}, \tag{2.141}$$

$$2A = \det \begin{pmatrix} 1 & x_{11}^{(e)} & x_{21}^{(e)} \\ 1 & x_{12}^{(e)} & x_{22}^{(e)} \\ 1 & x_{13}^{(e)} & x_{23}^{(e)} \end{pmatrix}. \tag{2.142}$$

The nodal index N is ordered in a counter clock-wise manner in the right-handed coordinates, and the symbol A represents the area of the simplex element. Figure 2.10 shows how the displacement field is interpolated using these shape functions. The simplex for three-dimension is left for readers as an exercise.

Exercise 2.15 Derive the shape functions for the three-dimensional simplex element (Fig. 2.8c).

2.5.4.2 Area and Volume Coordinates

Cartesian coordinates are not so convenient for the simplex elements. For 2-dimensional space, the area coordinates L_1, L_2, L_3 are defined as

$$x_1 = L_N x_{1N}^{(e)} = L_1 x_{11}^{(e)} + L_2 x_{12}^{(e)} + L_3 x_{13}^{(e)}$$
$$x_2 = L_N x_{2N}^{(e)} = L_1 x_{21}^{(e)} + L_2 x_{22}^{(e)} + L_3 x_{23}^{(e)}$$
$$1 = L_1 + L_2 + L_3. \tag{2.143}$$

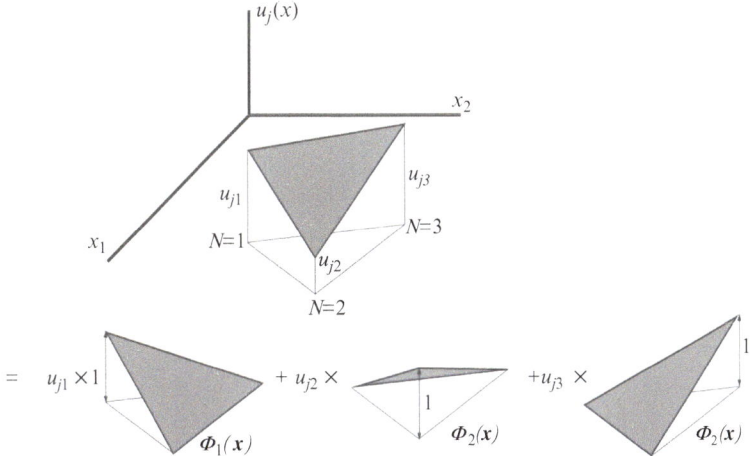

Fig. 2.10 Finite element approximation by two-dimensional simplex element

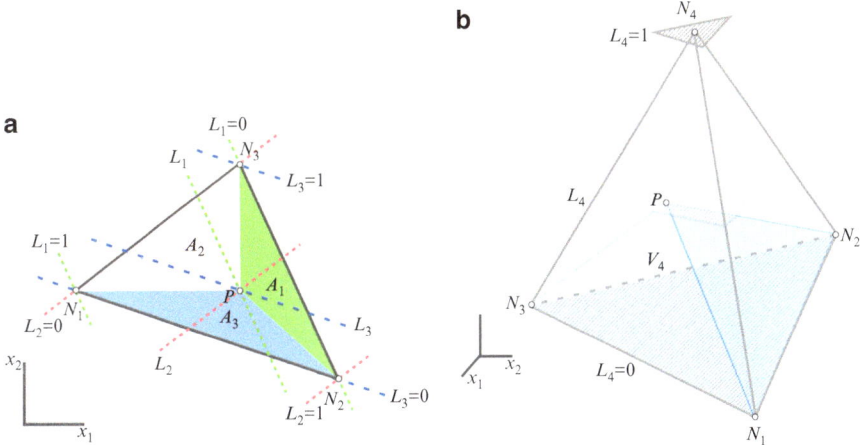

Fig. 2.11 Normalized coordinates. (**a**) Area coordinates and (**b**) volume coordinates

This definition uses the nodal coordinates $x_{iN}^{(e)}$ of a specific element, and the set of area coordinates is a local coordinates. As is shown in Fig. 2.11a, the area coordinate L_N is a normalized coordinate. It takes unit value at node N, zero on the opposite side, and constant value on the straight line parallel to the opposite side, i.e.

$$L_1 = \frac{\text{AreaP23}}{\text{Area } 123} = \frac{A_1}{A}, L_2 = \frac{\text{Area1P3}}{\text{Area } 123} = \frac{A_2}{A}, L_3 = \frac{\text{Area12P}}{\text{Area } 123} = \frac{A_3}{A}. \quad (2.144)$$

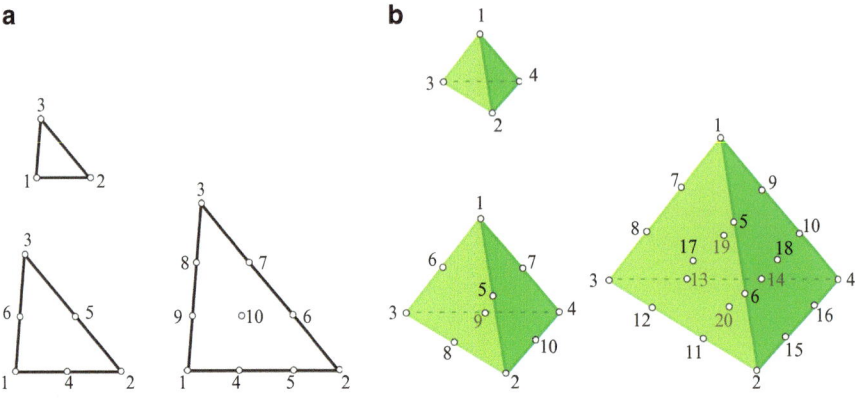

Fig. 2.12 Complex elements for (**a**) $k = 2$ and (**b**) $k = 3$

By inverting (2.143), it is found that the shape functions of 2-dimensional simplex, i.e. triangle element is

$$\Phi_N^{(e)} = L_N \tag{2.145}$$

For the 3-dimensional simplex of a tetrahedron (Fig. 2.8c), the volume coordinates L_1, L_2, L_3, L_4 are defined as

$$
\begin{aligned}
x_1 &= L_N x_{1N}^{(e)} = L_1 x_{11}^{(e)} + L_2 x_{12}^{(e)} + L_3 x_{13}^{(e)} + L_4 x_{14}^{(e)} \\
x_2 &= L_N x_{2N}^{(e)} = L_1 x_{21}^{(e)} + L_2 x_{22}^{(e)} + L_3 x_{23}^{(e)} + L_4 x_{24}^{(e)} \\
1 &= L_1 + L_2 + L_3 + L_4.
\end{aligned}
\tag{2.146}
$$

or

$$L_1 = \frac{\text{volume P234}}{\text{volume 1234}}, \quad L_2 = \frac{\text{volume 1P34}}{\text{volume 1234}}, \quad \text{and so on} \tag{2.147}$$

as illustrated in Fig. 2.11b. Again the shape functions are

$$\Phi_N^{(e)} = L_N \tag{2.148}$$

for the tetrahedral simplex element in three dimensions. Equation (2.145) or (2.148) satisfies the unit sum condition for shape functions, i.e. (2.99) in nature as is seen in the last part of (2.143) or (2.146).

Exercise 2.16 Examine (2.148).

2.5.4.3   Complex Elements

The k-dimensional simplex is the primitive object in the k-dimensional space, and simplex finite elements are convenient to discretize an arbitrarily shaped domain for analysis. However, it has only nodes used for interpolation at vertices of simplex primitive, and is not so good in accuracy of interpolation. Nodes are introduced for mid points on sides and faces for the higher order approximation with using the same element shape as simplex element (Fig. 2.12). Here $N_e > k + 1$ and such elements are referred to as a complex model. The shape functions for the quadratic triangle element of $k = 2$ and $N_e = 6$ are

$$\Phi_1^{(e)} = (2L_1 - 1)L_1, \ \Phi_2^{(e)} = (2L_2 - 1)L_2, \ \Phi_3^{(e)} = (2L_3 - 1)L_3$$

$$\Phi_4^{(e)} = 4L_1L_2, \ \Phi_5^{(e)} = 4L_2L_3, \ \Phi_6^{(e)} = 4L_3L_1 \tag{2.149}$$

and those for the quadratic tetrahedron element of $k = 3$ and $N_e = 10$ are

$$\Phi_1^{(e)} = (2L_1 - 1)L_1, \Phi_2^{(e)} = (2L_2 - 1)L_2, \ \Phi_3^{(e)} = (2L_3 - 1)L_3, \ \Phi_4^{(e)} = (2L_4 - 1)L_4,$$
$$\Phi_5^{(e)} = 4L_1L_2, \ \Phi_6^{(e)} = 4L_1L_3, \ \Phi_7^{(e)} = 4L_1L_4,$$
$$\Phi_8^{(e)} = 4L_2L_3, \ \Phi_9^{(e)} = 4L_3L_4, \ \Phi_{10}^{(e)} = 4L_2L_4.$$

$$\tag{2.150}$$

For integration of over triangular and tetrahedral domains, helpful are the formula

$$\int_{Triangle} \xi_1^p \xi_2^q \xi_3^r d\Omega = \frac{p!q!r!}{(p+q+r+2)!} 2A \tag{2.151}$$

and

$$\int_{Tetrahedron} \xi_1^p \xi_2^q \xi_3^r \xi_4^s d\Omega = \frac{p!q!r!s!}{(p+q+r+s+3)!} 6V. \tag{2.152}$$

Exercise 2.17 Examine the set of shape functions of (2.150) satisfy (2.99).

2.5.4.4   Multiplex Elements

Triangle and tetrahedron are the most fundamental primitive to divide an arbitrarily shaped domain of two- and three-dimensional space, respectively. Rectangle or rectangular parallelepiped is an alternative of a geometrical primitive, and can be

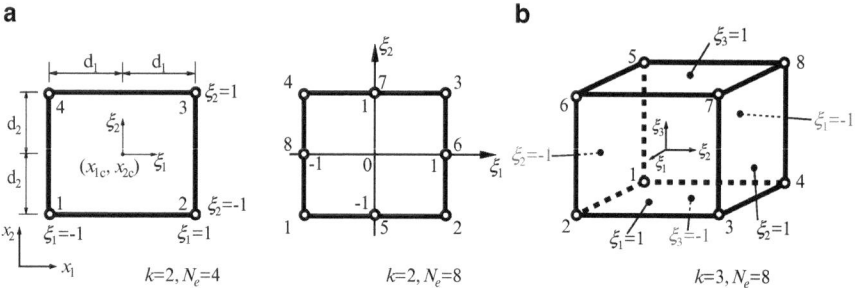

Fig. 2.13 Multiplex elements for (**a**) $k = 2$ and (**b**) $k = 3$

used as a shape of a finite element. The rectangle element is of $k = 2$ and $N_e = 4$, and the bilinear function

$$u_j(x_i) = a_0 + a_1 x_1 + a_2 x_2 + a_3 x_1 x_2. \tag{2.153}$$

is the displacement function. Although shape functions $\Phi_N^{(e)}$ are derived from the same procedure as (2.138)–(2.142) for the triangle element, the use of normalized coordinates (see Fig. 2.13)

$$\xi_j = \frac{1}{d_j}(x_j - x_{jc}) \text{ [not summed over } j] \tag{2.154}$$

gives us a shortcut to find shape functions which satisfy (2.99) by using that all the nodes at rectangular vertexes are $\xi_j = 1$ or -1. That is, the shape functions of rectangle element are

$$\Phi_1^{(e)} = \tfrac{1}{4}(1 - \xi_1)(1 - \xi_2), \quad \Phi_2^{(e)} = \tfrac{1}{4}(1 + \xi_1)(1 - \xi_2),$$
$$\Phi_3^{(e)} = \tfrac{1}{4}(1 + \xi_1)(1 + \xi_2), \quad \Phi_4^{(e)} = \tfrac{1}{4}(1 - \xi_1)(1 + \xi_2). \tag{2.155}$$

in normalized coordinates ξ_j, and substitution of (2.154) immediately derives the shape function in the original coordinates x_j.

The shape functions of rectangle of $k = 2$ and $N_e = 8$ with mid-side nodes are given as

$$\Phi_1^{(e)} = \tfrac{1}{4}(1 - \xi_1)(1 - \xi_2)(-\xi_1 - \xi_2 - 1), \quad \Phi_2^{(e)} = \tfrac{1}{4}(1 + \xi_1)(1 - \xi_2)(\xi_1 - \xi_2 - 1),$$
$$\Phi_3^{(e)} = \tfrac{1}{4}(1 + \xi_1)(1 + \xi_2)(\xi_1 + \xi_2 - 1), \quad \Phi_4^{(e)} = \tfrac{1}{4}(1 - \xi_1)(1 + \xi_2)(-\xi_1 + \xi_2 - 1),$$
$$\Phi_5^{(e)} = \tfrac{1}{2}(1 - \xi_1^2)(1 - \xi_2), \quad \Phi_6^{(e)} = \tfrac{1}{2}(1 + \xi_1)(1 - \xi_2^2),$$
$$\Phi_7^{(e)} = \tfrac{1}{2}(1 - \xi_1^2)(1 + \xi_2), \quad \Phi_8^{(e)} = \tfrac{1}{2}(1 - \xi_1)(1 - \xi_2^2).$$

$$\tag{2.156}$$

For the rectangular parallelepiped of $k = 3$ and $N_e = 8$, the shape functions are

$$\Phi_1^{(e)} = \tfrac{1}{8}(1 - \xi_1)(1 - \xi_2)(1 - \xi_3), \quad \Phi_2^{(e)} = \tfrac{1}{4}(1 + \xi_1)(1 - \xi_2)(1 - \xi_3),$$
$$\Phi_3^{(e)} = \tfrac{1}{4}(1 + \xi_1)(1 + \xi_2)(1 - \xi_3), \quad \Phi_4^{(e)} = \tfrac{1}{4}(1 - \xi_1)(1 + \xi_2)(1 - \xi_3),$$
$$\Phi_5^{(e)} = \tfrac{1}{2}(1 - \xi_1)(1 - \xi_2)(1 + \xi_3), \quad \Phi_6^{(e)} = \tfrac{1}{2}(1 + \xi_1)(1 - \xi_2)(1 + \xi_3),$$
$$\Phi_7^{(e)} = \tfrac{1}{2}(1 - \xi_1)(1 + \xi_2)(1 + \xi_3), \quad \Phi_8^{(e)} = \tfrac{1}{2}(1 - \xi_1)(1 - \xi_2)(1 + \xi_3). \quad (2.157)$$

These elements are called Serendipity family.

Exercise 2.18 Examine the sets of shape functions given by (2.156) and (2.157) satisfy (2.29).

2.5.5   Shape Functions: Isoparametric Elements

The simplex in one-, two- and three-dimensional spaces is a line segment, triangle and tetrahedron, and those are composed of straight line and flat plane segments. Biological organs however have a geometrical shape of free-form in general and it is not always convenient to reconstruct it with simplexes of straight edges. Use of finite elements with curved shape is a way to improve the approximation of the shape of an arbitrary geometry. The curved element is represented by a mapping of a one-to-one transformation between any global coordinate system x_j and local coordinate system ξ_i,

$$x_j = x_j(\xi_i). \tag{2.158}[18]$$

The exact mapping is not available for a free-formed object, and a general alternative is a parametric mapping for example with polynomials-based transformation functions.

The transformation equation (2.154) used for multiplex elements is the simplest case of such a transformation from a square in ξ_i to a rectangle x_j. First explained is the extension for the transformation to a quadrilateral in global coordinate, then to a curved quadrilateral. For a quadrilateral element shown in Fig. 2.14a, the transformation polynomial is a bilinear function and is written with using the nodal coordinates in the global coordinates $x_{iN}^{(e)}$ for $N_e = 4$ as

$$x_i(\boldsymbol{\xi}) = a_0 + a_1\xi_1 + a_2\xi_2 + a_3\xi_1\xi$$
$$= \Phi_N^{(e)}(\boldsymbol{\xi})x_{iN}^{(e)} \tag{2.159}$$

---

[18] We assume the existence of the inverse mapping $\xi_i = \xi_i(x_j)$.

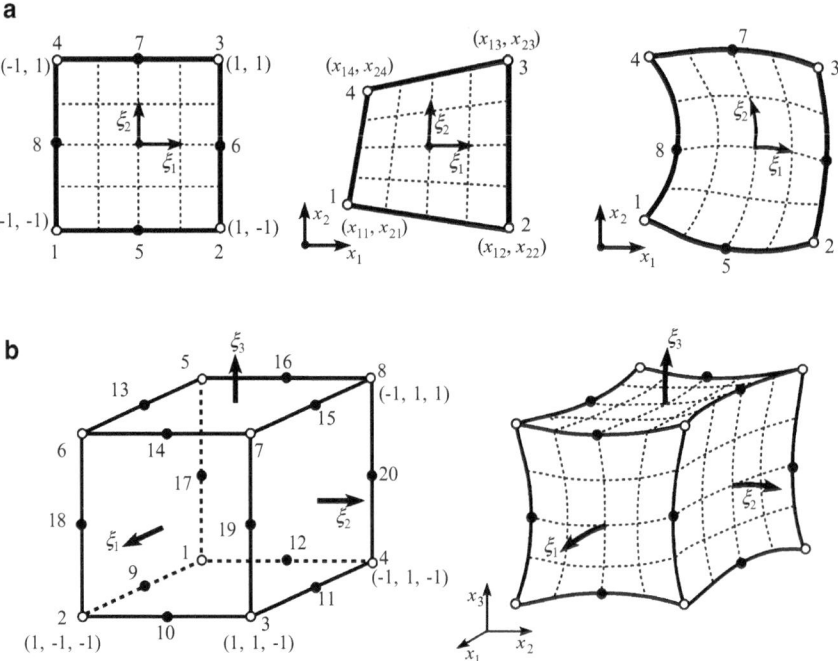

Fig. 2.14 Isoparametric elements for (**a**) $k = 2$ and (**b**) $k = 3$

where $\Phi_N^{(e)}(\boldsymbol{\xi})$ are identical to the shape functions of (2.155) for displacement interpolation (2.153) of $u_{iN}^{(e)}$. For a curvilinear quadrilateral element, four corner nodes are not sufficient and mid-side nodes are recruited such that $N_e = 8$. The shape functions in (2.156) are used to define the mapping from a square to a curvilinear quadrilateral. The parametric expression for coordinate transformation is the same as those for displacement field interpolation, and these curved elements are called isoparamteric elements. In general, for isoparametric elements, we have

$$u_i(\mathbf{x}) = \Phi_N^{(e)}(\mathbf{x})u_{iN}^{(e)} \qquad (2.98)$$

$$x_i(\boldsymbol{\xi}) = \Phi_N^{(e)}(\boldsymbol{\xi})x_{iN}^{(e)} . \qquad (2.159)$$

Since the shape functions are described in the normalized local coordinates ξ_i, the derivatives of a shape function with respect to global coordinates x_i are expressed by using the chain rule

$$\frac{\partial \Phi_N^{(e)}}{\partial x_i} = \frac{\partial \Phi_N^{(e)}}{\partial \xi_j} \frac{\partial \xi_j}{\partial x_i} \qquad (2.160)$$

where $\frac{\partial \xi_j}{\partial x_i}$ is calculated by inverting Jacobian

$$\frac{\partial x_i}{\partial \xi_j} = \frac{\partial \Phi_N^{(e)}}{\partial \xi_j} x_{iN}^{(e)} \tag{2.161}$$

The integrant for instance for element stiffness is thus given in local coordinates variable ξ_i and the integral is carried out in the local coordinate system by paying attention to the difference of domain scale with determinant of Jacobian,

$$dx_1 dx_2 dx_3 = \det \left| \frac{\partial x_i}{\partial \xi_j} \right| d\xi_1 d\xi_2 d\xi_3 \tag{2.162}$$

In practice, the integral is done numerically by the Gauss quadrature for instance

$$\int_{\Omega(\xi)} \Lambda(\xi_i) d\xi_1 d\xi_2 d\xi_3 = \sum_I w_I \Lambda(\xi_{iI}) \tag{2.163}$$

where ξ_{iI} and w_I is a set of coordinates of integration points and their weight.

For further details on finite element method, textbooks by Zienkiewicz et al. (2005), Hughes (2003), Oden (2000) and many others are helpful.

Exercise 2.19 Consider a curvilinear hexahedron element of $N_e = 20$ shown in Fig. 2.14b. The shape function for a corner node $N = 1$ and a mid-side noted $N = 9$ are given as

$$\begin{aligned}
\Phi_1^{(e)} &= \tfrac{1}{8}(1 - \xi_1)(1 - \xi_2)(1 - \xi_3)(-\xi_1 - \xi_2 - \xi_3 - 2) \\
\Phi_9^{(e)} &= \tfrac{1}{4}(1 - \xi_1{}^2)(1 - \xi_2)(1 - \xi_3)
\end{aligned} \tag{2.164}$$

Confirm them and find the full set of shape functions.

2.6   Computational Biomechanics Problems

2.6.1   Lattice Continuum Modeling for Cancellous Bone Structure

Elastic modulus of cancellous bone is characterized by an exponential function of density or volume fraction in Sect. 2.2.2. This mainly comes from experimental observations and much effort has been devoted to identify the exponent. This is also examined by a theoretical consideration as cell structure (Gibson 1985).

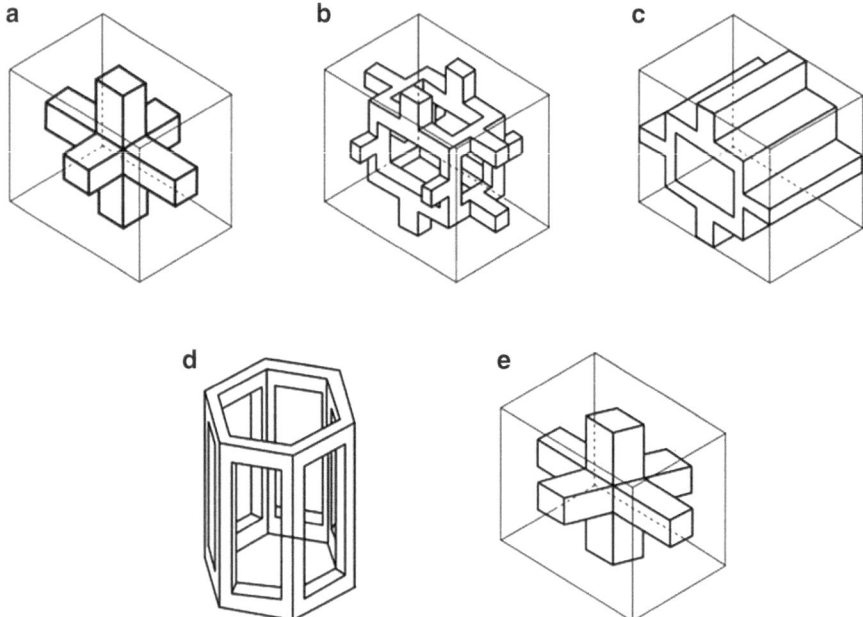

Fig. 2.15 Typical unit cells for structural model of trabecular architecture

The structural anisotropy of trabecular architecture in cancellous bone is not taken into account in the elastic modulus.

The trabecular architecture has a characteristic structure in accordance with its mechanical function. The majority is rod- or beam-like shaped trabeculae in the region of a low volume fraction and is plate-like trabeculae in the region of a high volume density. Its characteristics orientation is well known in a famous work by von Meyer (1856, 1867, 2011; Skedros and Brand (2011)), and Wolff (1892, 1986). A cell structure model is a way of investigating the mechanical characteristics of trabecular architecture. There are several cell models that characterize the geometrical structure of cancellous bone made of rod and plate elements (Fig. 2.15).

According to the Wolff's law, the cell of orthogonally intersected rods (Fig 2.15a) is a basic model (Adachi et al. 1999). The mechanical conditions of this cell structure are illustrated in Fig. 2.16 where each rod element is subjected to an axial force $N_{ij}(i = j)$ and shear forces $N_{ij}(i \neq j)$ and a torsional moment $M_{ij}(i = j)$ and bending moments $M_{ij}(i \neq j)$. The rod and beam theory gives us the strain tensor ε_{ij} and the curvature tensor κ_{ij} of the unit cell as[19]

---

[19] Summention convention is not applied for equations with $(\ )^{\#}$ in its equation number.

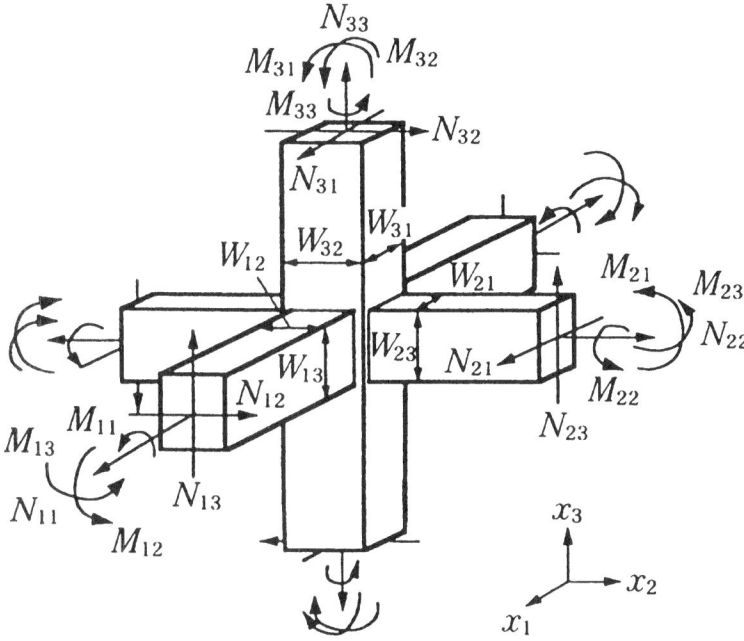

Fig. 2.16 Forces and moments working on a unit cell of orthogonally intersected lattice structure

$$\varepsilon_{ii} = \frac{N_{ii}}{E_i A_i}, \ \varepsilon_{ij} = \frac{1}{24}\left(\frac{L_i^2 N_{ij}}{E_i I_{ij}} + \frac{L_j^2 N_{ji}}{E_j I_{ji}}\right),$$

$$\kappa_{ii} = \frac{M_{ii}}{G_i J_i}, \ \kappa_{ij} = \frac{M_{ij}}{G_i I_{ik}}, \tag{2.165}^{\#}$$

where E_i and G_i are the longitudinal and shear elastic moduli, J_i and I_{ik} are the polar second moment and second moment of cross sectional area with respect to x_k axis of rod element of cross-sectional area A_i aligned in x_i axis. Denoting the size of unit cell by L_i, the macroscopic stress tensor T_{ij} and couple stress tensor μ_{ij} are defined as an average over the area of cell face as

$$T_{ii} = \frac{N_{ii}}{L_j L_k}, \ T_{ij} = \frac{N_{ij}}{L_j L_k}, \ \mu_{ii} = \frac{M_{ii}}{L_j L_k}, \ \mu_{ij} = \frac{M_{ij}}{L_j L_k} \tag{2.166}^{\#}$$

In the context of the couple stress theory, a skew symmetric part of stress tensor does not contribute to an internal energy and the mean part of couple stress tensor is indefinite. Thus, using the symmetric part σ_{ij} of stress components and the deviatric part m_{ij} of the couple stress

$$\sigma_{ij} = \sigma_{ji} = \frac{1}{2}(T_{ij} + T_{ji}) \text{ and } m_{ij} = \mu_{ij} - \frac{1}{3}\delta_{ij}\mu_{kk}, \tag{2.167}^{\#}$$

the constitutive equation is given as

$$\sigma_{ii} = E_{ii}^* \varepsilon_{ii}, \; \sigma_{ij} = 2G_{ij}^* \varepsilon_{ij},$$
$$m_{ii} = 4G_{ii}^* L_{ii}^{*2} \kappa_{ii}, \; m_{ij} = 4G_{ik}^* L_{ik}^{*2} \kappa_{ij}. \tag{2.168}^{\#}$$

where E_{ij}^*, G_{ik}^* and L_{ik}^* are apparent constants of a continuum given as

$$E_{ii}^* = E_i S_i, \; G_{ii}^* = \frac{G_i J_i}{L_i^2 L_j L_k},$$

$$L_{ii}^* = \frac{J_i}{2}, \; L_{ij}^* = \frac{L_i}{48^{1/2}} \left(1 + \frac{E_i L_j I_{ij}}{E_j L_i I_{ji}}\right)^{1/2}$$

$$S_i = \frac{A_i}{L_j L_k} = \eta_{ij} \eta_{ik}, \; A_i = W_{ij} W_{ik} \text{ and } \eta_{ij} = \frac{W_{ij}}{L_j}. \tag{2.169}^{\#}$$

Here W_{ij} is the width of element. It is noted here the homogenized mechanical property of cell structures as a continuum is dependent on mechanical properties as well as the characteristic size of a constitutive element of the cell structure.

Summarizing the constitutive equation (2.168) in a general form

$$\sigma_{ij} = C_{ijkl} \varepsilon_{kl} \text{ and } m_{ij} = D_{ijkl} \kappa_{kl} \tag{2.170}$$

in the coordinate system x_i, an elasticity constant tensor C_{ijkl}' in a different coordinate system x_i' is written by

$$C_{ijkl}' = \theta_{i'i} \theta_{j'j} \theta_{k'k} \theta_{l'l} C_{ijkl} \tag{2.171}$$

using the transformation tensor $\theta_{i'i}$. The elasticity of uniaxial tensile test is the reciprocal of the component D_{1111}' of compliance tensor D_{ijkl}'

$$D_{ijkl}' C_{klmn}' = \frac{1}{2}(\delta_{im}\delta_{jn} + \delta_{in}\delta_{jm}). \tag{2.172}$$

As the most basic case of $E_i = E$, $L_i = L$ and $W_{ij} = W_{ik} = W_i$, a normalized elastic constant is

$$\frac{\tilde{E}(\theta_{1i})}{E} = \left[\frac{\theta_{11}^4}{\eta_2 \eta_3} + \frac{\theta_{12}^4}{\eta_1 \eta_3} + \frac{\theta_{13}^4}{\eta_1 \eta_2} + \theta_{11}^2 \theta_{12}^2 \left(\frac{\eta_1^3 + \eta_2^3}{\eta_3 \eta_1^3 \eta_2^3}\right) \right.$$
$$\left. + \theta_{12}^2 \theta_{13}^2 \left(\frac{\eta_2^3 + \eta_3^3}{\eta_1 \eta_2^3 \eta_3^3}\right) + \theta_{13}^2 \theta_{11}^2 \left(\frac{\eta_3^3 + \eta_2^3}{\eta_2 \eta_3^3 \eta_1^3}\right)\right]^{-1}. \tag{2.173}$$

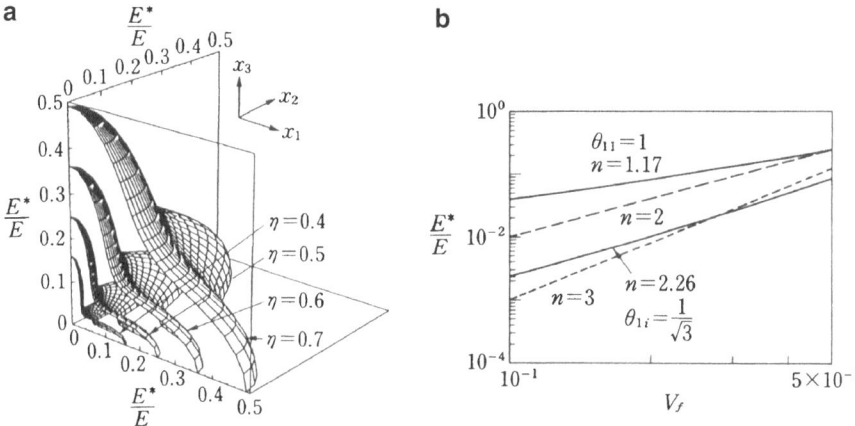

Fig. 2.17 Anisotropy and volume fraction dependent elastic properties

Figure 2.17a shows the polar plot of normalized elastic modulus as the distance from the origin with respect to the tensile orientation. This illustrates the anisotropic characteristics of a trabecular network by means of cell structure.

When the density and elastic constant of the material are the same for all rod elements of the cell structure, the apparent density is proportional to the volume fraction. The exponent model thus gives the apparent elastic modulus as

$$\hat{E} = EV_f^n. \tag{2.174}$$

For the cell structure of $\eta_i = \eta$, the volume fraction is

$$V_f = 3\eta^2 - 2\eta^3 \tag{2.175}$$

and (2.173) is reduced to

$$\frac{\tilde{E}(\theta_{1i})}{E} = \left[\frac{\theta_{11}^4 + \theta_{12}^4 + \theta_{13}^4}{\eta^2} + \frac{\theta_{11}^2\theta_{12}^2 + \theta_{12}^2\theta_{13}^2 + \theta_{13}^2\theta_{11}^2}{\eta^4} \right]^{-1}. \tag{2.176}$$

The apparent elastic modulus becomes the minimum for the direction of $\theta_{11} = \theta_{12} = \theta_{13} = 1/\sqrt{3}$, and the maximum for the direction of $\theta_{11} = 1$, $\theta_{12} = \theta_{13} = 0$. The regression lines for these maximum and minimum with respect to the volume fraction result in the exponent $n = 1.17$ for the maximum and $n = 2.26$ for the minimum as is shown in Fig. 2.17b. This is another aspect of cell structures model representing the relationship of elastic modulus and volume fraction or apparent density.

2.6.2 Lateral Deformation Analysis of Spinal Column

Spine system is the largest subsystem of a skeletal structure, and is composed of the vertebral column and rib cages. The vertebral column has seven cervical vertebrae, twelve thoracic vertebrae, five lumber vertebrae and sacrum, and supported by many ligaments including anterior and posterior longitudinal ligaments. The rib cage consists of the sternum, costae and other components. The vertebal column exhibits a normal posterior curvature in the thoracic region and the normal anterior curvature in the lumber region. Abnormal curvatures include the excessive posterior curvature (kyphosis or so-called roundback), the excessive anterior curvature (lordosis or so-called swayback/saddle back) and the lateral curvature (scoliosis). The scoliosis is the most common abnormal curvatures, more common in females. Especially, the idiopathic scoliosis frequently initiates and progresses during a rapid growth period of adolescence. The growth force, thus, has been suggested as one of primary causes of the idiopathic scoliosis as well as an aggregative factor for its progress although this is recognized as a multi-factorial disorder. Computational biomechanics analysis is expected as a powerful tool to examine the mechanical initiation and progress of lateral deformation (e.g., Stokes and Laible 1990; Kawabata et al. 1988; Tadano et al. 1996; Azegami et al. 1998; Villemure et al. 2002, 2004; Aoyama et al. 2008; Shi et al. 2011).

A three-dimensional finite element model (Todoh et al. 2001a, b) of the spinal system of vertebral column and rib cage is shown in Fig. 2.18. A spinal column system with rib cages is developed in three-dimensional finite elements based on the normal geometry and anatomy. The vertebrae and intervertebral discs reconstruct a physiological curve in a sagittal plane. Facet joints connect adjacent vertebrae at their posterior elements. The sternum is connected via costal cartilage and costae to the column at costovertebral joints. These bony components are interconnected by anterior and posterior longitudinal ligaments, intertransverse ligaments, supra-spinous and interspinous ligaments, capsular ligament and yellow ligament. The vertebrae and intervertebral disks are composed of 8-node brick elements primarily for cortical and cancellous bone as well as the annulus fibrosus and nucleus pulposus. The ligaments are represented by spar elements of one-dimensional tension material. All the components are assumed to be linear elastic body as is shown in Table 2.1 in reference to literature (Goel et al. 1994; Maurel et al. 1997; Azegami et al. 1998). Though nonlinearity is intrinsic in soft tissues deformation, the linear elasticity has been extensively used for scoliosis study especially for the etiology because the initiation of lateral curve occurs within small strain.

The growth strain method is recruited to impose growth on components of the spinal system. By referring to volumetric growth of vertebral bodies in adolescence, a growth strain of 0.5% is considered for vertebrae in an isotropic manner for an increment, with which 6-month growth is represented by ten increments. As an extremum situation, no growth is assumed for the ligaments, while the 0.5% growth strain is assumed for all components except ligaments. The spinal column is fixed to the base at the distal end of L5 vertebral body. Lack of symmetry with respect to the

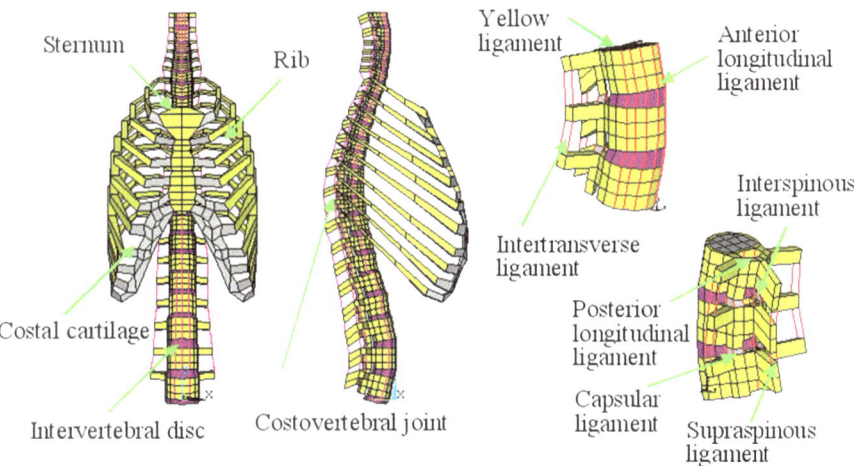

Fig. 2.18   Three-dimensional finite element model of spinal column with rib cage

Table 2.1   Material properties

Components	Tissue	E (MPa)	υ	A (mm^2)
Vertebral body	Cortical bone	17,000	0.3	
	Cancellous bone	200	0.3	
Rib and sternum	Cortical bone	17,000	0.3	
Facet joint	Articular tissue	10,000	0.4	
Intervertebral disc		2.5	0.3	
Costal cartilage	Cartilage	500	0.3	
Anterior longitudinal ligament	Ligament	10		0.5
Posterior longitudinal ligament	Ligament	20		0.5
Capsular ligament	Ligament	20		1.2
Yellow ligament	Ligament	50		0.4
Intertransverse ligament	Ligament	10		3.6
Interspinous ligament	Ligament	3		3.0
Supraspinous ligament	Ligament	3		5.0

Source: Taken from Goel et al. (1994), Maurel et al. (1997) and Azegami et al. (1998)

sagittal plane is considered for an internal pressure in rib cage due to the asymmetric arrangement of a cardiac and arterial system, and a small disturbance is introduced at T8 level. The scoliosis as well as the lordosis/kyphosis and rotation are computed based on accumulation of internal stress/strain caused by the unbalanced volumetric growth between hard and soft tissue components.

Spinal curve in Fig. 2.19 was induced by no growth in ligament inconsistent with the normal growth in other components. The curve is of rightward convex thoracic pattern with the maximum deflection at T8 level. It showed a significant vertebral axial rotation toward the convex side of the scoliotic deflection, although little change was caused in the curve in the sagittal plane. This curve has the basic

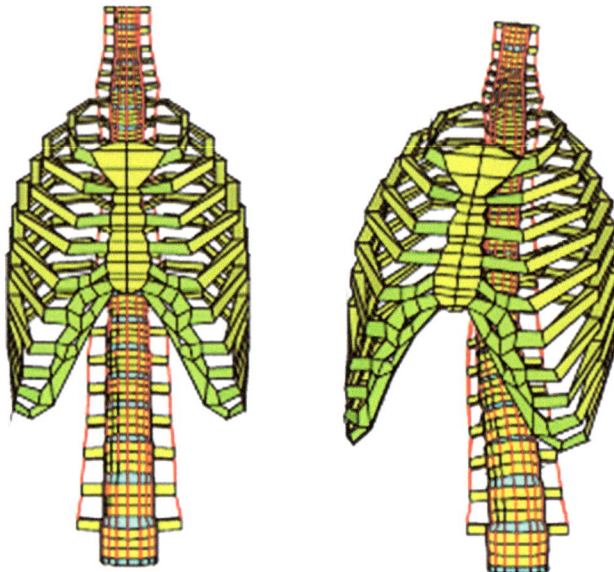

Fig. 2.19 Frontal view of spinal system under balanced growth (*left*) and unbalanced growth (*right*) conditions between hard and soft tissue components

features found in typical thoracic scoliotic curves of scoliosis observed clinically. The transverse sectional geometry in a thoracic level exhibited reasonable vertebral displacement and rotation in the rib cage in comparison with those in the geometry observed clinically in transverse CT images (Fig. 2.20). The right and left rib cage space has no clear difference in this analysis, although the left half space of the rib cage is usually lager than the right half one. The deformity of internal organs in the rib cage was not taken into account in this analysis. However, the asymmetric arrangement of lung and cardiovascular system could be one of primal causes of such an asymmetric rib cage deformation in scoliosis.

The growth strain of homogeneous volume rate induces only the proportional increase in size in the entire spinal system without any internal stress or strain in it. This is an ideal situation for balanced growth between the components. In this analysis, the inconsistent growth among the structural components in spinal system was considered, and it was found such a growth unbalance might cause scoliotic deformations although the etiology of scoliosis is multifactorial. This is a computational biomechanics consideration for animal experiments in which over-constraints at the posterior ligamentous sites resulted in severe scoliotic deformation (Kasuga 1994) and for a theoretical concept that a combination of coronal plane asymmetry and posterior structure tightening initiates the spinal rotation (Dickson et al. 1984).

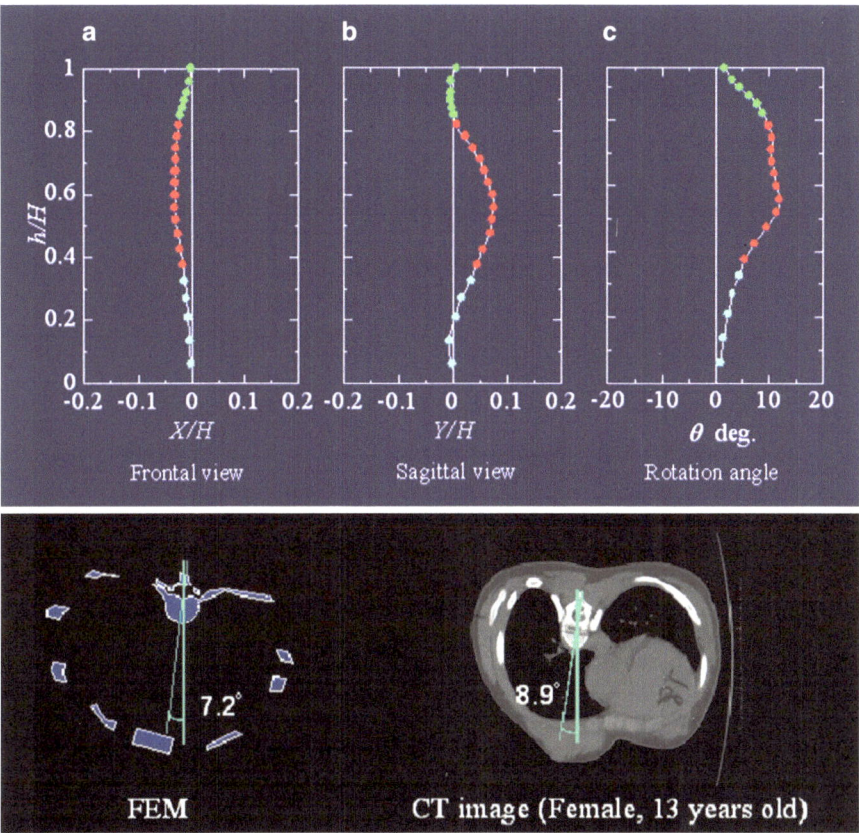

Fig. 2.20 Deflection/rotation of spinal column (*top*) and horizontal section at thoracic vertebra T8 (*bottom*) under unbalanced growth condition

2.6.3   Stress Analysis of Temporomandibular Joint Disc

The temporomandibular joint (TMJ) is an articulation joint between the condyle of mandibular bone and the mandibular fossa of temporal bone, a part of skull. It is responsible for talking and masticating and is used in highly frequently in daily life. Mandibular condyle has the geometry similar to spheroid in which the major axis approximately is oriented in a mediolateral direction. The articular surface of temporal bone is divided into two parts of the posterior concave glenoid fossa and the anterior convex articular eminence, and the joint motion is rather complex in comparison with skeletal joints of upper and lower limbs. The joint motion has two different modes of rotation and translation. The mandular condyle remains within the articular surface of concave fossa during usual speech and chewing with a small mouth opening, while, occasionally, the condyle translationally travels toward an anterior direction beyond the eminence such as gaping and yawning which accompany a large mouth opening. Such articular motion is supported by the temporomandibular disc traveling together with the condyle along the temporal bone surface. Internal derangement frequently accompanies the dislocation of articular disc toward the anterior or anteromedical directions that is known as anterior disc displacement. It happen to be accompanied by disc perforation. The biomechanical irregularity is considered as one of the major causes of disc derangement, and computational stress analyses have been conducted (e.g. Chen and Xu 1994; Tanaka et al. 1994; Beek et al. 2000, Tanaka et al. 2003, Pérez del Palomar and Doblaré 2006; Tanaka et al. 2008).

Bruxism such as clenching or grinding is reported to be associated with TMJ disorder as a factor causing lateral pterygoid muscle tonus, anterior disc displacement and so on. For the computational analysis of a continuous clenching particularly considered as one of the major factors inducing disorder, the three-dimensional geometry data of each TMJ component of soft and hard tissues were obtained by using oblique MRI sagittal slices of a healthy subject to construct the finite element model of TMJ consisted of articular disc, mandibular fossa, condyle, and connective tissue (Fig. 2.21a). Bone components were treated as rigid bodies and soft tissue components were modeled as the Kelvin-type[20] viscoelastic continuum (Kelvin solid[21]). To impose loading conditions during clenching, the masseter muscle, the temporal muscle, the medio pterygoid muscle and the superior head of lateral pterygoid muscle (SLPM) were modeled based on the cephalometric radiographs of the same subject. It is assumed that the positional relationship between the maxilla and the mandible remained unchanged at the first molar and the motion of the central point of anterior teeth was limited to the median plane during clenching by considering the symmetrical clenching at dens molaris. To consider these conditions, the muscle origins, the bottom face of the condyle, the first molar, and the central point of the anterior teeth were connected with the rigid link representing the mandible (Fig. 2.21c).

The material properties of soft tissues were defined as shown in Table 2.2 where $E_0 = E_R \tau_\sigma / \tau_\varepsilon$ and E_R are the instantaneous and relaxed elastic moduli, and τ_σ and τ_ε are relaxation constants for constant stress and strain (see footnote 21). To treat bone

---

[20] Viscoelastic model of Kelvin-type is a parallel connection of a serial connection of a spring of constant μ_1 and a dashpot (damping component) of constant η and another spring of constant μ_2. The force-elongation relationship is written as $F + \tau_u \dot{F} = K_R(u + \tau_F \dot{u})$ where $K_R = \mu_2, \tau_u = \eta/\mu_1$ and $\tau_F = (\eta/\mu_2)[1 + \mu_2/\mu_1]$. The elongation $c(t)$ for the force of a unit-step function $1(t)$ and the force $k(t)$ for an elongation $1(t)$ are $c(t) = K_R^{-1}[1 - (1 - \tau_u/\tau_F) \exp(-t/\tau_F)]1(t)$ and $k(t) = K_R[1 - (1 - \tau_F/\tau_u) \exp(-t/\tau_u)]1(t)$, respectively. These $c(t)$ and $k(t)$ are called the creep function and relaxation function.

[21] Kelvin-type continua is an extension of Kelvin-type viscoelastic model of springs and dashpot. Force and elongation are replaced with stress and strain as $\sigma + \tau_\varepsilon \dot{\sigma} = E_R(\varepsilon + \tau_\sigma \dot{\varepsilon})$ where $E_R = \mu_2, \tau_\varepsilon = \eta/\mu_1$ and $\tau_\sigma = (\eta/\mu_2)[1 + \mu_2/\mu_1]$. This is the model of one-dimensional Kelvin-type continuum, and stress–strain relationship is given by $\sigma(t) = \int_0^t k(t - \tau)\dot{\varepsilon}(\tau)d\tau$ or $\varepsilon(t) = \int_0^t c(t - \tau)\dot{\sigma}(\tau)d\tau$. This model is extended to three dimensions in the similar manner in one-dimensional elasticity to three-dimensional one, establishing the Kelvin-type continuum (Kelvin solid, also known as standard linear solid). It is noted here, for the Kelvin solid, the constants μ_1 and μ_2 are elastic moduli with the unit of Pa and η is viscosity with the unit of Pa · s.

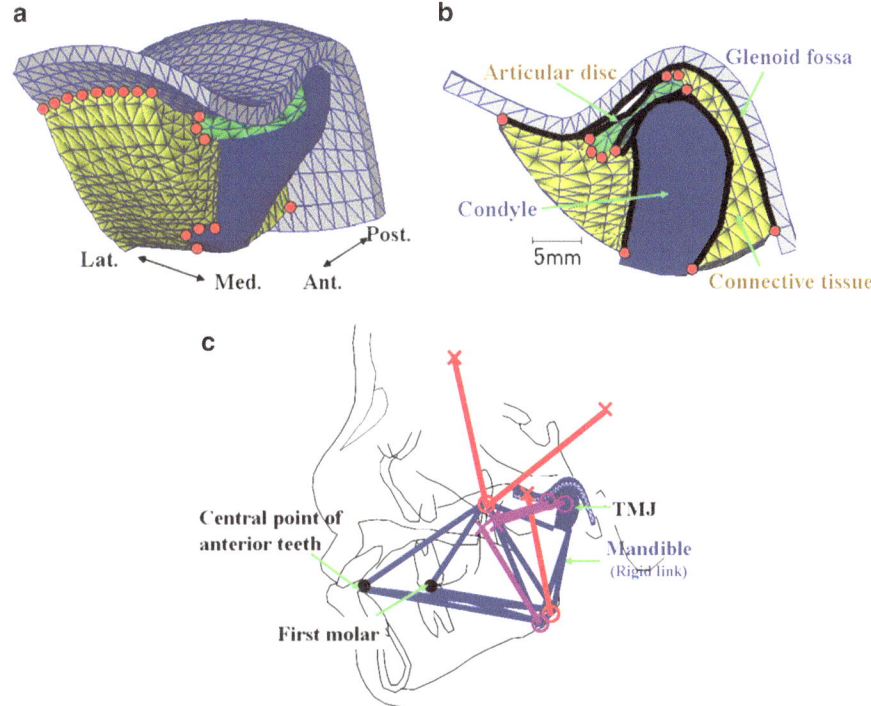

Fig. 2.21 Finite element model of temporomandibular joint. (**a**) Finite element TMJ model. Medial half of connective tissue is not shown. (**b**) Sagittal section of TMJ model. (**c**) Rigid link of the mandible with muscle spar element

Table 2.2 Material properties of articular disc and connective tissue

	E_0 (MPa)	E_R (MPa)	τ_ε (s)	ν
Articular disc	30.9	15.8	31.2	0.4
Connective tissue	1.54	0.21	11.6	0.4

Source: Taken from Tanaka et al. (1999, 2002)

components as rigid bodies, a sufficiently large elastic modulus was given to them. Contact elements were arranged at the bone–soft tissue interface as shown by thick lines in Fig. 2.21b to take account of the contacting condition, and the coefficient of a friction of $\mu = 10^{-3}$ was used. The articular surface of fossa and the muscle origins were fixed on the coordinate system, and the relative motion of mandibular condyle was considered with respect to the mandibular fossa. Based on the assumption described above, the first molar was fixed on the coordinate system, and the displacements and rotations of the central point of the anterior teeth were constrained within the median plane. Furthermore, we assumed that each muscle force was proportional to the physiological cross-section of the muscle and that a half of SLPM was attached to the disc and the rest to the fovea pterygoidea. Each muscle force was determined so that the condyle displacement at the beginning of clenching (time=0 s) was equal to the condyle displacement defined by Beek et al. (2000) in their clenching analysis, and was applied for 10 min.

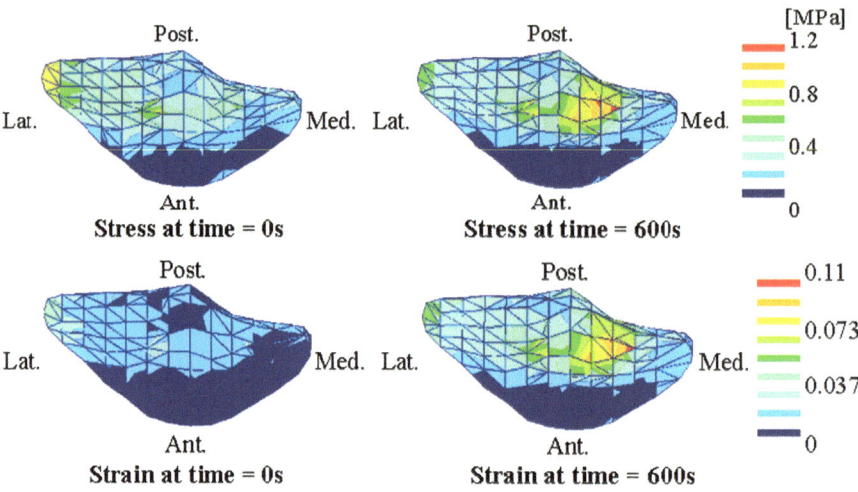

Fig. 2.22 Equivalent stress and strain in the disc during prolonged clenching

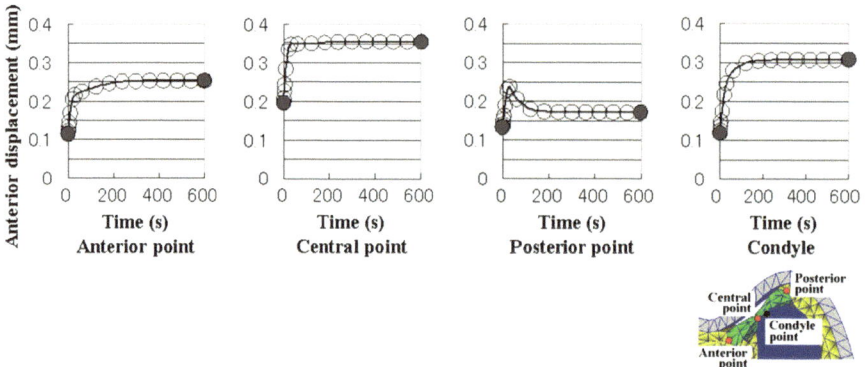

Fig. 2.23 Anterior displacement of disc points and of condylar surface point during prolonged clenching

In the disc, the equivalent stress (von Mises stress) and strain were relatively high in the posterior band and elevated with according to the continuation of clenching (Fig. 2.22). In the connective tissue surrounding the disc, a relatively high stress/strain was observed mainly in the retrodiscal region. In particular, the retrodiscal tissue strain almost tripled at the end of the continuous clenching (time $=600$ s). This result indicates that the mechanical condition of soft tissue is moderate at the initial contact and becomes severer with time. In addition, the anterior displacement of anterior point of the disc increased with time and kept as is throughout the clenching period, while that of posterior point increased first and then decreased (Fig. 2.23). The continuous clenching enforces the disc to move toward the anterior direction, possibly resulting in damage or degeneration of the retrodiscal tissue and permanent anterior disc displacement accordingly. Further computational analyses are found in literatures (e.g. Hirose et al. 2006; Tanaka et al. 2008; Mori et al. 2010).

2.6.4   Deformation Analysis of Cornea: Inverse Problems for Natural Shape and Intraocular Pressure

The cornea is a transparent component located at the most frontal part of the eye. The cornea refracts lights creating an image on the retina. The focus of cornea is fixed while a lens adjusts a focus depending on a distance to an object. Nevertheless the cornea is very important for a good vision because approximately two-thirds of optical power is accounted for by the cornea. The geometrical shape of cornea such as a curvature and its imperfection are critical features to determine vision irregularities of myopia (short-sightedness), hyperopia (long-sightedness), astigmatism and so forth. The geometry of cornea in vivo is dependent on the material properties of cornea tissue and the intraocular pressure of the fluid inside of the eye. That is, important are the cornea shape deformed under the continuously working intraocular pressure and resulted from refractive surgeries. Thus, computational analyses of cornea deformation have been conducted for modeling of the mechanical properties, the refractive surgery, the tonometry and so forth (e.g. Hanna et al. 1988, 1989; Vito and Carnell 1992; Bryton and McDonnel 1996; Alastrue et al. 2006; Elsheikh and Anderson 2005; Elsheikh et al. 2009; Tanaka et al. 2007; Elsheikh and Wang 2007; Guzman et al. 2011).

The material properties and intraocular pressure are crucial for the deformation analysis in general. The cornea shape at a natural state[22] in absence of intraocular pressure is the bases for deformation analysis, but a measured cornea shape is deformed one in the presence of intraocular pressure. Concerning the intraocular pressure, the Goldmann applanation tonometry has been recognized as a golden standard in several different types of tonometry. However, it does not consider the effects of cornea geometry such as thickness and curvature and of material properties although those are common variations in individual eyes and affects the pressure value obtained by applanation tonometry (e.g. Liu and Roberts 2005; Velten et al. 2006; Elsheikh et al. 2006). Therefore, identification (estimation) of the natural shape of cornea and the intraocular pressure is inevitable for computational deformation analyses as well as the identification of material properties. These problems are categorized in inverse or back analyses.

---

[22] The original state of solid body in which no loads are applied is called the natural state or stress-free configuration. It is assumed zero stress and zero strain in the solid body. In the case of cornea, the in vivo state is a loaded state and the natural state will appear when the intraocular pressure is removed.

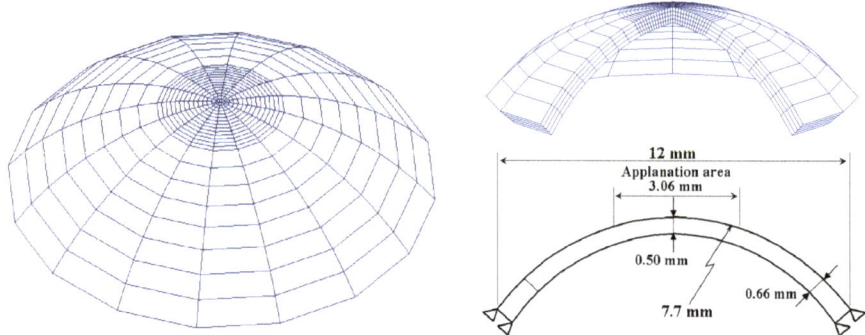

Fig. 2.24 Corneal model. Three-dimensional finite element model (*left*), its cut image (*right top*) and dimensions in cross section (*right bottom*)

A three-dimensional finite element model used here is shown in Fig. 2.24. The identification problems of a natural shape and an intraocular pressure are described separately. The strain energy function of cornea has much varieties in the past reports, i.e. linear elasticity/nonlinear hyper elasticity and isotropy/transverse isotropy/orthotropy and others. Here, a hyperelastic solid body of transversely isotropy is assumed. The isotropic contribution is a second order form of (2.72), the anisotropic contribution is for transversely isotropy (Weiss et al. 1996), and the volumetric deformation is given in the context of quasi-incompressibility $J = \sqrt{I_3} \approx 1$.

$$
\begin{aligned}
\Psi_{isoch}^{iso} &= C_{10}(\hat{I}_1 - 3) + C_{20}(\hat{I}_1 - 3)^2, \\
\Psi_{isoch}^{aniso} &= C_4(\exp(\hat{I}_4 - 1) - \hat{I}_4), \quad \Psi_{vol} = \frac{\kappa}{2}(\sqrt{I_3} - 1)
\end{aligned}
\tag{2.177}
$$

The constants of strain energy function are determined based on the inflation test (Anderson et al. 2004) as $C_{10} = 0.033$ MPa, $C_{20} = 16.5$ MPa, $C_4 = 0.50$MPa and $\kappa = 16.5$MPa. Denoting the position of a material point of cornea at the natural state by X, the cornea shape in vivo is written as $\mathbf{x} = X + \mathbf{u}$ with displacement $\mathbf{u}(X)$ under the intraocular pressure p. When the position of a surface point is obtained as $\mathbf{x}^*$ on the surface Γ_s for shape measurement, a difference between the measured position and the computed position

$$
E_X(X) = \int_{\Gamma_s} (\mathbf{x}^* - (X + \mathbf{u}(X)))^T (\mathbf{x}^* - (X + \mathbf{u}(X))) d\Gamma
\tag{2.178}
$$

works as the shape error function which is minimized for the identification of the natural shape $\mathbf{X}$. This is a nonlinear optimization of the shape determination problem studied in the field of optimum structural design and solved here by the

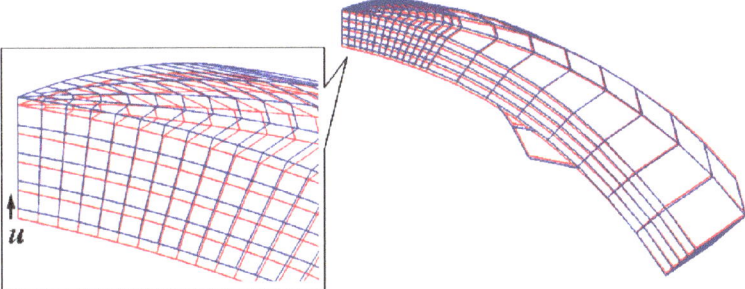

Fig. 2.25 Identified natural shape of cornea (*red*) from deformed shape (*blue*) under intraocular pressure. Only a quarter is shown

traction method (Azegami 1994). Figure 2.25 shows the identified natural shape of cornea (profile illustrated in red) when the deformed shape (profile illustrated in blue) is measured for the intraocular pressure of 14 mmHg. In this case, the anterior displacement along the cornea axis due to the pressure is approximately 15% of the central thickness of cornea and its effects on the surface geometry of cornea are not negligibly small. The anterior displacement is not uniform along the axis resulting in a compressive strain in cornea tissue. It is noted here that the superposition principle is not valid for nonlinear problem and the influence of deformation due to the intraocular pressure has a significant meaning in the further analyses such as for surgical intervention starting from an in vivo stress/strain condition.

The second is the identification problem of intraocular pressure. The Imbert–Fick law is the theoretical base for the standard method of Goldmann applanation tonometry, and does not take account of corneal geometry and properties. For the natural shape of cornea estimated, the intraocular pressure p is applied first and then the applanation deformation is enforced for the central zone of 3.06 mm in diameter on the outer surface of the cornea. That is, the same level for the axial position $x_3 = X_3 + u_3$ is enforced under the axial compression for all the surface points of the applanation zone, and the applanation force F is calculated as a resultant of axial reactions in the zone. The pressure that is quotient of an applanation force divided by an applanated area by the Imbert–Fick law is not identical to the pressure assumed for the computational analysis. In the case of the model of Fig. 2.25, the applanation force results in a quotient pressure approximately 20% different from the pressure applied for the cornea model. The pressure identification problem is thus needed and becomes a one-dimensional minimization of the error between the applanation force F^* experimentally observed and the force $F(p)$ calculated by a computational analysis when the natural geometry and the material properties are available.

Both of natural shape and material properties of cornea are unknown in a real situation, and thus the simultaneous identification is a problem to be considered. However, for such a problem, a conventional single point data of one pair of

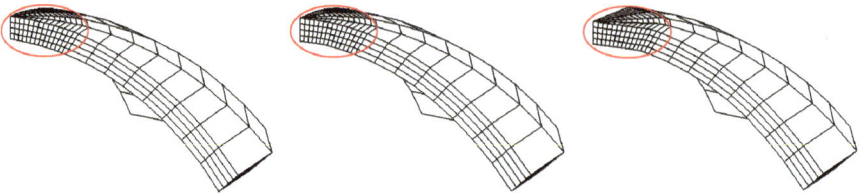

Fig. 2.26 Computational analysis of continuous applanation for model-based applanation force–area relationship. Cornea deformation with radius of contact area of 0 (*left*), 0.765 mm (*center*) and 1.53 mm (*right*)

applanation force and area is not sufficient for the identification of multiple parameters of material constants and pressure. This simultaneous identification problem becomes well-defined and is solved based on a continuous applanation analysis such as in Fig. 2.26, if an applanation force-applanation area (or displacement) curve is available (Tanaka et al. 2009). This kind of model-based computational analyses has a capability of suggesting a scheme for an advanced experimental observation in biomedicine.

2.6.5   Stress Analysis of Proximal Femur: Image-Based Analysis and Simulation

The proximal femur is a representative target in orthopaedic bone mechanics due to its characteristic trabecular architecture pointed by von Meyer (1856), Wolff (1892) and others. Many computational stress analyses and remodeling simulation studies dealt with this bone structure. The progress of biomedical imaging such as X-ray computed tomography (CT) imaging enabled us to examine the details of bone structure with various image resolutions at hospitals in practice, research laboratories and advanced synchrotron radiation facilities. Another progress in high performance computing provides tremendous computer powers at a super computer center and also high computing powers at laboratories and personal desk-sides. Computational biomechanics analysis is thus extended to an image-based modeling for mechanics analysis. Especially, a two- or three-dimensional medical image composed of pixels or voxels is used in geometrical modeling for biomechanics analysis with pixel-based rectangle or voxel-based brick elements. These give the straight direction of image-based two- or three-dimensional pixel/voxel finite element analyses.

Figure 2.27a shows a CT image of proximal femur of a subject at hospital use devices, with image resolution of 700 μm. Each voxel is used to define an eight nodes brick element for the finite element analysis. An isotropic linear elastic body is assumed for bone element, and the elastic modulus E is defined for each element as

$$E = 0.012\rho_{HA} + 0.25 \ \text{(GPa)} \tag{2.179}$$

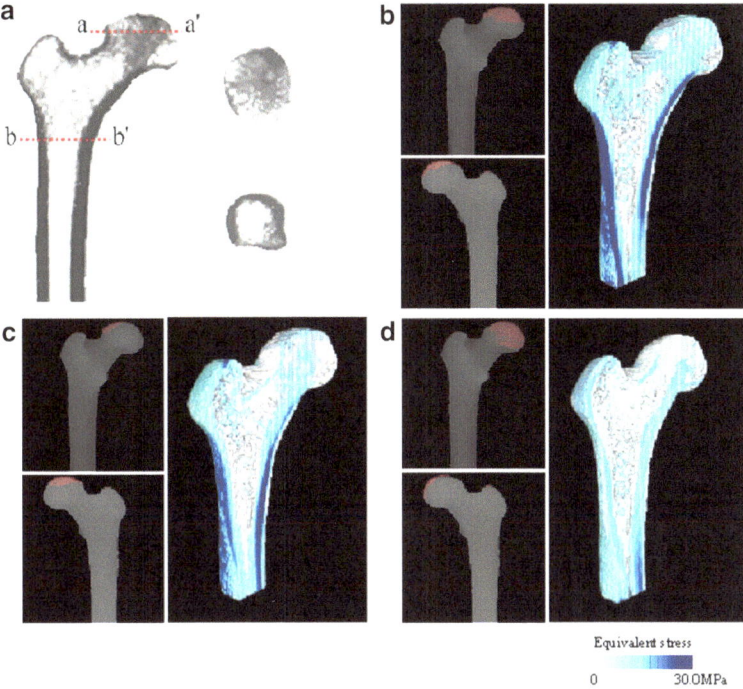

Fig. 2.27 Image-based stress analysis of proximal femur. (**a**) CT image, and the equivalent stress distribution under (**b**) adductive, (**c**) neutral and (**d**) adductive loading condition those are illustrated by *pink color* in left half of each figure

based on the density ρ_{HA} (calcium hydroxyapatite in mg/cm^3) that is converted from the CT value referring to those of CT phantom

$$CT = 1.1316\rho_{HA} + 3.2383. \tag{2.180}$$

Poisson's ratio is 0.3. The loads working at the femoral head and great trochanter are determined by referring to the body weight of the subject and the force distribution pattern representing the neutral, adductive and adductive positions (Beaupré et al. 1990). The image resolution used here is not sufficient to examine detail of each trabecular but is sufficient to examine the local distribution of apparent density of cancellous bone. As seen in the figure, the stress distribution is deeply related to the bone density distribution and is very dependent on the loading conditions. This is subject-specific information, and the image-based computational biomechanics analyses provide us quantitative mechanical information on skeletal functions for diagnosis, while medical images directly give us much information on skeletal geometry in detail.

Characteristics of trabecular architecture are recognized as the result of mechanical adaptation by bone remodeling, because it is regulated by mechanical stimuli in part

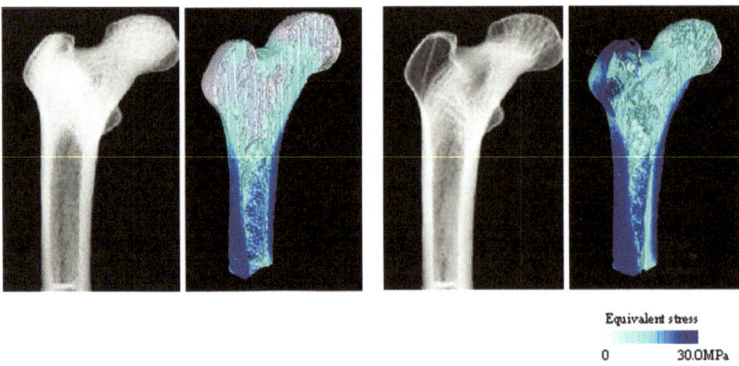

Fig. 2.28 Internal structure simulated healthy (*left*) and osteoporotic (*right*) femur with the same outer configuration, and equivalent stress distribution under the same body weight condition

while it is essentially a physiological process regulated by activities of osteoclasts and osteoblasts. There are efforts to analyze the bone remodeling computationally (e.g. Cowin 1986; Carter 1987; Turner 1992; Mullender et al. 1994; Adachi et al. 1997; Ruimerman et al. 2005; Tsubota et al. 2009; Kwon et al. 2010a), although each of them is not mentioned here. Relevant to the stress condition of proximal femur, the simulated internal structures are shown in Fig. 2.28 for healthy and disuse-mediated osteoporotic cases with the same outer surface configuration of a femur of Fig. 2.27 (Kwon et al. 2010b). The simulated internal structure for a healthy case exhibited a good correlation between its local volume fraction and the bone density by CT images, and that for osteoporotic case reproduced a typical bone loss pattern characterized by grade 4 in Singh index (Singh et al. 1970) loosing trabeculae inside of the great trochanter, reducing the presence of principal tensile trabecular group and accompanying the enlargement of Wards triangle. The same loading conditions are enforced to these two different internal structures and the equivalent stress distribution is analyzed and shown in Fig. 2.28. A significant decrease of cancellous bone volume resulted in a higher stress distribution in these region compared with the healthy case, increasing the fracture risk in femoral head and neck. These findings are not qualitative but quantitative for an individual model by subject specific medical images. Stress analyses for simulated bone structures may provide insights into a long term change in the bone structure by remodeling and its mechanical condition after treatment such as implantation in coming future.

2.7   Summary

This chapter starts with tiny fundamentals of solid mechanics of linear and nonlinear elastic bodies applicable for biomechanical analysis of biosolids' behaviors, and constitutive models for some of representative biological solid bodies are explained. A core of finite element method is then described for small strain linear elastic and finite strain nonlinear hyperelastic bodies based on the variational principles of minimum potential energy and virtual work. In these days, many general purpose finite element packages are commercially available and it is easy to access computer analyses of solid mechanics including biomechanics problems. However, it is important to have theoretical backgrounds even for a package software user, because a rational and reasonable biomechanical modeling and result interpretation are possible only for users who know what the software is doing or can do, especially for the problems coming from biology and biomedicine. Several solid biomechanics problems are demonstrated. Some of them are categorized in a direct or forward analysis where all the fundamentals required for the analysis are available a priori and the response of a target solid body is a problem. On the contrary, some problems involve the identification process using a known behavior of biosolid body first to analyze the unknown behavior. Such inverse analysis problems are inevitable in the biomechanical problems in vivo, especially for subject- and patient-specific analysis. The major field of computational biomechanics is the analysis of tissue/organ behaviours at this moment. However, the application field of computational biomechanics analysis is extending for or being shifted towards the prediction oriented analysis/simulations by incorporating adaptively changing capability of biosolids, since it is the essential characteristics that distinguish biomechanics problems from conventional engineering problems.

References

Adachi T, Tomita Y, Sakaue H, Tanaka M (1997) Simulation of trabecular surface remodeling based on local stress nonuniformity. JSME Int J Ser C 40(4):782–892

Adachi T, Tomiya Y, Tanaka M (1999) Three-dimensional lattice continuum model of cancellous bone for structural and remodeling simulation. JSME Int J Ser C 42(3):470–480

Alastrue V, Calvo B, Pena E, Doblare M (2006) Biomechanical modeling of refractive corneal surgery. J Biomech Eng 128(1):150–160

Anderson K, El-Sheikh A, Newson T (2004) Application of structural analysis to the mechanical behaviour of the cornea. J R Soc Interface 1(1):1–13

Aoyama T, Azegami H, Kawakami N (2008) Nonlinear buckling analysis for etiological study of idiopathic scoliosis. J Biomech Sci Eng 3(3):399–410

Ashman RB, Cowin SC, van Buskirk WC, Rice JC (1984) A continuous wave technique for the measurement of the elastic properties of cortical bone. J Biomech 17(5):349–361

Austman RL, Milner JS, Holdsworth DW, Dunning CE (2008) The effect of the density-modulus relationship selected to apply material properties in a finite element model of long bone. J Biomech 41(15):3171–3176

Azegami H (1994) Solution to domain optimization problems. Trans Jpn Soc Mech Eng Ser A 60:1479–1485

Azegami H, Murachi S, Kitoh J, Ishida Y, Kawakami N, Makino M (1998) Etiology of idiopathic scoliosis: computational study. Clin Orthop Relat Res 357:229–236

Beaupré GS, Orr TE, Carter DR (1990) An approach for time-dependent bone modeling-application: a preliminary remodeling simulation. J Orthop Res 8:662–670

Beek M, Koolstra JH, van Ruijven LJ, van Eijden TMGJ (2000) Three-dimensional finite element analysis of the human temporomandibular joint disc. J Biomech 33:307–316

Bryton M, McDonnel P (1996) Constitutive laws for biomechanical modeling of refractive surgery. J Biomech Eng 118(4):473–481

Buskirk WCV, Cowin SC, Ward RN (1981) Ultrasonic measurement of orthotropic elastic constants of bovine femoral bone. Trans ASME J Biomech Eng 103:67–72

Carter DR (1987) Mechanical loading history and skeletal biology. J Biomech 20:1095–1109

Carter DR, Hayes WC (1976) Bone compressive strength: the influence of density and strain rate. Science 196(4270):1174–1176

Carter DR, Hayes WC (1977) The compressive behaviour of bone as a two-phase porous structure. J Bone Joint Surg 59A:954–962

Chen J, Xu L (1994) A finite element analysis of the human temporomandibular joint. J Biomech Eng 116:401–407

Ciarelli MJ, Glodstein SA, Kuhn JL, Cody DD, Brown MB (1991) Evaluation of orthogonal mechanical properties and density of human trabecular bone form the major metaphyseal regions with materials testing and computed tomography. J Orthop Res 9:674–682

Cilingir AC, Ucar V, Kazan R (2007) Three-dimensional anatomic finite element modeling of hemi arthroplasty of human hip joint. Trends Biomater Artif Organs 21(1):63–72

Cowin SC (1983) The mechanical and stress adaptive properties of bone. Ann Biomed Eng 3 (3–4):263–295

Cowin SC (1986) Wolff's law of trabecular architecture at remodeling equilibrium. J Biomech Eng 108:83–88

Cowin SC (ed) (1989) Bone mechanics. CRC Press, Boca Raton

Cowin SC (ed) (2001) Bone mechanics handbook, 2nd edn. CRC Press, Boca Raton

Currey JD (2002) Bones: structure and mechanics. Princeton University Press, Princeton

Dickson RA, Lawton JO, Archer IA, Butt WP (1984) The pathogenesis of idiopathic scoliosis. Biplanar spinal asymmetry. J Bone Joint Surg Br 66-B(1):8–15

Duchemin L, Bousson V, Raossanaly C, Bergot C, Laredo JD, Skalli W, Mitton D (2008) Prediction of mechanical properties of cortical bone by quantitative computed tomography. Med Eng Phys 30(3):321–328

Elsheikh A, Anderson K (2005) Comparative study of corneal strip extensometry and inflation tests. J R Soc Interface 2(3):177–185

Elsheikh A, Wang D (2007) Numerical modelling of corned biomechanical behaviour. Comp Meth Biomech Biomed Eng 10:85–95

Elsheikh A, Wang D, Kotecha A, Brown M, Garway-Heath D (2006) Evaluation of Goldmann applanation tonometru using a nonlinear finite element ocular model. Ann Biomed Eng 34 (10):1628–1640

Elsheikh A, Ross S, Alhasso D, Rama P (2009) Numerical study of the effect of corneal layered structure on ocular biomechanics. Curr Eye Res 34(1):26–35

Flynn C, Taberner A, Nielsen P (2011) Modeling the mechanical response of in vivo human skin under a rich set of deformations. Ann Biomed Eng 39(7):1935–1946

Fung YC, Fronek K, Patitucci P (1979) Pseudoelasticity of arteries and the choice of its mathematical expression. Am J Physiol- Heart 237(5):H620–H631

Gasser TC, Ogden RW, Halzapfel GA (2005) Hyperelastic modelling of arterial layers with distributed collagen fibre orientations. J R Soc Interface 3(6):15–35

Gibson LJ (1985) The mechanical behaviour of cancellous bone. J Biomech 18(5):317–328

Goel VK, Park H, Kong W (1994) Investigation of vibration characteristics of the ligamentous lumbar spine using the finite element approach. J Biomech Eng 116(4):377–383

Guzmán FA, Castilla AA, Guarnieri FA and Rodríguez FR (2011) Intraocular pressure: Goldmann tonometry, computational model, and calibration equation. J Glaucoma [Epub ahead of print] doi: 10.1097/IJG.0b013e31822f4747

Hanna K, Jouve F, Bercovier M, Waring G (1988) Computer simulation of lamellar keratectomy and myopic keratomeleusis. J Refract Surg 4(6):222–231

Hanna K, Jouve F, Waring G, Ciarlet P (1989) Computer simulation of arcuate and radial incisions involving the corneo-scleral limbus. Eye 3:227–239

Hernandez CJ, Beaupre GS, Keller TS, Carter DR (2001) The influence of bone volume fraction and ash fraction on bone strength and modulus. Bone 29(1):74–78

Hirose M, Tanaka E, Tanaka M, Fujita R, Kuroda Y, Yamano E, van Eijden TMGJ, Tanne K (2006) Three-dimensional finite-element model of human temporomandibular joint disc during prolonged clenching. Eur J Oral Sci 114(5):441–448

Hodgskinson R, Currey JD (1992) Young's modulus, density and material properties in cancellous bone over a large density range. J Mater Sci Mater Med 3(5):377–381

Holzapfel GA (2000) Nonlinear solid mechanics: a continuum approach for engineering. Wiley, Chichester

Holzapfel GA, Weizacker HW (1998) Biomechanical behavior of the arterial wall and its numerical characterization. Comput Biol Med 28:377–392

Holzapfel GA, Casser TC, Ogden RW (2000) A new constitutive framework for arterial wall mechanics and a comparative study of material model. J Elasticity 61(1–3):1–48

Hughes TJR (2003) The finite element method–linear static and dynamic finite element analysis. Dover, Mineola

Kasuga K (1994) Study on scoliosis experimentally caused by binding posterior spinous processes in rats. J Jpn Orthop Assoc 68:798–807

Kawabata H, Ono K, Seguchi Y, Tanaka M (1988) Idiopathic scoliosis and growth -a biomechanical consideration. J Jpn Orthop Assoc 62:167–170

Knets I (2002) Peculiarities of the structure and mechanical properties of biological tissue. Meccanica 37(4–5):375–384

Knets I, Malmeisters A (1977) Deformability and strength of human compact bone tissue. In: Brankov G (ed) Mechanics of biological solids: Proceedings of Euromech Colloquium 68, Bulgarian Academy of Sciences, Sofia, pp 133–141

Kwon J, Naito H, Matsumoto T, Tanaka M (2010a) Simulation model of trabecular bone remodeling considering effects of osteocyte apoptosis and targeted remodeling. J Biomech Sci Eng 5(5):539–551

Kwon J, Naito H, Matsumoto T, Tanaka M (2010b) Computational study on trabecular bone remodeling in human femur under reduced weight-bearing conditions. J Biomech Sci Eng 5 (5):552–564

Lanir Y (1987) Skin mechanics In: Skalak R, Chien S (eds) Handbook of bioengineering. McGraw-Hill, New York, pp 11.11–11.25

Laville A, Laporte S, Skalli W (2009) Parametric and subject-specific finite element modelling of the lower cervical spine. Influence of geometrical parameters on the motion patterns. J Biomech 42(10):1409–1415

Liu J, Roberts CJ (2005) Influence of corneal biomechanical properties on intraocular pressure measurement; quantitative analysis. J Cataract Refract Surg 31(1):146–155

Lotz JC, Gerhart TN, Hayes WC (1990) Mechanical properties of trabecular bone from the proximal femur: a quantitative CT study. J Comput Assist Tomogr 14(1):107–114

Lotz JC, Gerhart TN, Hayes WC (1991) Mechanical properties of metaphyseal bone in the proximal femur. J Biomech 24(5):317–329

Martin RB, Burr DB, Sharkey NA (1998) Skeletal tissue mechanics. Springer, New York

Maurel N, Lavaste F, Skalli W (1997) A three-dimensional parameterized finite element model of the lower cervical spine, study of the influence of the posterior articular facets. J Biomech 30 (9):921–931

Morgan EF, Bayraktar HH, Keaveny TM (2003) Trabecular bone modulus-density relationships dependent on anatomic site. J Biomech 36:897–904

Mori H, Horiuchi S, Nishimura S, Nikawa H, Murayama T, Ueda K, Ogawa D, Kuroda S, Kawano F, Naito H, Tanaka M, Koolstra JH, Tanaka E (2010) Three-dimensional finite elemen analysis of cartilaginous tissues in human temporomandibular joint during prolonged clenching. Arch Oral Biol 55:879–886

Mullender MG, Huiskes R, Weinans H (1994) A physiological approach to the simulation of bone remodeling as a self-organizational control process. J Biomech 27(11):1389–1394

Muller R, Puegsegger PR (1996) Analysis of mechanical properties of cancellous bone under conditions of simulated bone atrophy. J Biomech 29(8):1053–1060

Oden JT (2000) Finite element of nonlinear continua. Dover, Mineola (originally published in 1972 by Mc-Grow Hill, New York)

Oden JT, Reddy JN (1976) Variational methods in theoretical mechanics. Springer, Heidelberg

Pandolfi A, Holzapfel GA (2008) Three-dimensional modeling and computational analysis of the human cornea considering distributed collagen fibril orientations. J Biomech Eng 130(6):061006

Pérez del Palomar A, Doblaré M (2006) Finite element analysis of the temporomandibular joint during lateral excursions of the mandibule. J Biomech 39(12):2153–2163

Reilly DT, Burstein AH (1975) The elastic and ultimate properties of compact bone tissue. J Biomech 8(6):393–405

Rho JY, Hobato MC, Ashman RB (1995) Relations of mechanical properties to density and CT numbers in human bone. Med Eng Phys 17(5):347–355

Rice JC, Cowin SC, Bowman JA (1988) On the deoendence of the elasticity and strength of cancellous bone on apparent density. J Biomech 21(2):155–168

Roy A, Dupps WJ Jr (2011) Patient-specific modeling of corneal refractive surgery outcomes and inverse estimation of elastic property changes. J Biomech Eng 133(1):011002

Ruimerman R, Hillbers P, van Rietbergen B, Huiskes R (2005) A theoretical framework for strain-related trabecular bone maintenance and adaptation. J Biomech 38(4):931–941

Shi L, Wang D, Driscoll M, Villemure I, Chu WC, Cheng JC, Aubin CE (2011) Biomechanical analysis and modeling of different vertebral growth patterns in adolescent idiopathic scoliosis and healthy subjects. Scoliosis 23(6):11

Singh M, Nagrath AR, Maini PS (1970) Changes in trabecular pattern of the upper end of the femur as an index of osteoporosis. J Bone Joint Surg Am 52:457–467

Skedros JG, Brand RA (2011) Biographical sketch: Georg Hermann von Meyer (1815–1892). Clin Orthop Relat Res 469(11):3072–3076

Stokes IAF, Laible JP (1990) Three-dimensional osseo-ligamentous model of the thorax representing initiation of scoliosis by asymmetric growth. J Biomech 23(6):589–596

Studer H, Larrea X, Riedwyl H, Buchler P (2009) Biomechanical model of human cornea based on stromal microstructure. J Biomech 43(5):836–842

Taber LA (2004) Nonlinear theory of elasticity: applications in biomechanics. World Scientific, Singapore

Tadano S, Kanayama M, Ukai T, Kaneda K (1996) Morphological modeling and growth simulation of idiopathic scoliosis. In: Hayashi K, Ishikawa H (eds) Computational biomechanics. Springer, Tokyo, pp 67–88

Tanaka E, Tanne K, Sakuda M (1994) A three-dimensional finite element model of the mandible including the TMJ and its application to stress analysis in the TMJ during clenching. Med Eng Phys 16:316–322

Tanaka E, Tanaka M, Miyawaki K, Tanne K (1999) Viscoelastic properties of canine temporomandibular joint disc in compressive load-relaxation. Arch Oral Biol 44:1021–1026

Tanaka E, del Pozo R, Sugiyama M, Tanne K (2002) Biomechanical response of retrodiscal tissue in the temporomandibular joint under compression. J Oral Maxillofac Surg 60:546–551

Tanaka M, Tanaka E, Tadoh M, Asai D, Kuroda Y (2003) Stress analysis of anterior-disc-displaced temporomandibular joint using individual finite element model. JSME Int J Ser C 46(4):1400–1408

Tanaka M, Matsumoto T, Naito H, Jinno T (2007) Shape optimization for corneal refractive surgery planning realizing structural profile by trimming operation. Seventh World Congress on Structural and Multidisciplinary Optimization, Seoul, Korea, 21–25 May 2007, pp 625–630

Tanaka E, Hirose M, Koolstra JH, van Eijden TMGJ, Fujita R, Tanaka M, Tanne K (2008) Modeling of the effect of friction in the temporomandibular joint on displacement of its disc during prolonged clenching. J Oral Maxillofac Surg 66(3):462–468

Tanaka M, Matsumoto T, Naito H, Tanaka H (2009) Identification method of subject specific corneal material constants and intraocular pressure. IV International Congress on Computational Bioengineering, p 109

Todoh M, Tanaka M, Ebara S (2001a) Idiopathic scoliosis analysis by finite element model of spine: influence of spinal ligament constraint. Jpn J Clin Biomech 22:233–236

Todoh M, Tanaka M, Ebara S (2001b) Analysis of idiopathic scoliosis by finite element method with unbalanced growth: study on human spine model with spinal ligaments. Spinal Deformity J Jpn Scoliosis Soc 16:5–8

Tong P, Fung YC (1976) The stress–strain relationship for the skin. J Biomech 9(10):649–657

Tsubota K, Suzuki T, Yamada T, Hojo M, Makinouchi M, Adachi T (2009) Computer simulation of trabecular remodeling in human proximal femur using large-scale voxel FE models: approach to understanding Wolfff's law. J Biomech 42:1088–1094

Turner CH (1992) On Wolff's law of trabecular architecture. J Biomech 25(1):1–9

Turner CH, Cowin SC (1987) Dependence of elastic constants of an anisotropic porous material upon porosity and fabric. J Mater Sci 22:3178–3184

Turner CH, Rho J, Kanano Y, Tsui TY, Pharr GM (1999) The elastic properties of trabecular and cortical bone tissues are similar: results from two microscopic measurement technique. J Biomech 32(4):437–441

Vaishnav RH, Young JT, Patel DJ (1973) Distribution of stresses and of strain-energy density through the wall thickness in a canine aortic segment. Circ Res 32:577–583

van Rietbergen B, Majumdar S, Pistoia W, Newitt DC, Kothari M, Laib A, Ruegsegger PR (1998) Assessment of cancellous bone mechanical properties form micro-FE models based on micro-CT, pQCT and MR images. Technol Health Care 6(5–6):4132–4420

Velten K, Gunther M, Oberacher-Velten I, Lorenz B (2006) Finite-element simulation of corneal applanation. J Cataract Refract Surg 32(7):1073–1074

Villemure I, Aubin CE, Dansereau J, Labelle H (2002) Simulation of progressive deformities in adolescent idiopathic scoliosis using a biomechanical model integrating vertebral growth modulation. J Biomech Eng 124(6):784–790

Villemure I, Aubin CE, Dansereau J, Labelle H (2004) Biomechanical simulations of the spine deformation process in adolescent idiopathic scoliosis from different pathogenesis hypotheses. Eur Spine J 13:83–90

Vito RP, Carnell PH (1992) Finite element based mechanical models of the cornea for pressure and indenter loading. Refract Corneal Surg 8(2):146–151

von Meyer GH (1856) Lehrbuch der physiologischen anatomie des menschen. Verlag von Wilhelm Engelmann, Leipzig

von Meyer GH (1867) Die architectur der spongiosa. Reichert und Du Bois-Reymond's Archiv 8:615–628

von Meyer GH (2011) The classic: the architecture of the trabecular bone (Tenth contribution on the mechanics of the human skeletal framework), (Die Architectur der Spongiosa) trans: Brand RA. Clin Orthop Relat Res 469(11):3079–3084

Wang C, Garcia M, Lu X, Lanir Y, Kassab GS (2006) Three-dimensional mechanical properties of porcine coronary arteries: a validated two-layer model. Am J Physiol Heart Circ Physiol 291: H1200–H1209

Washizu K (1982) Variational method for elasticity and plasticity, 3rd edn. Pergamon Press, Oxford

Weiss JA, Maker BN, Govindjee S (1996) Finite element implementation of incompressible, transversely isotropic hyperelasticity. Comput Methods Appl Mech Eng 135(1–2):107–128

Wolff J (1892) Das gesetz der transformation der knochen. Verlag von August Hirschwald, Berlin

Wolff J (1986) The law of bone remodelling (trans: Marquet P, Furlong R). Springer, Berlin

Wolff J (2010) The classic: on the inner architecture of bones and its importance for bone growth 1870. Clin Orthop Relat Res 468:1056–1065

Yang G, Kabel J, van Rietbergen B, Odgaard A, Huiskes R, Cowin SC (1999) The anisotropic Hooke's law for cancellous bone and wood. J Elast 53(2):125–146

Zienkiewicz OC, Taylor RL, Zhu JZ (2005) The finite element method: its basis and fundamentals, 6th edn. Elsevier Butterworth-Heinemann, Burlington

Zysset PK, Guo XE, Hoffer CE, Moore KE, Goldstein SA (1999) Elastic modulus and hardness of cortical and trabecular bone lamellae measured by nanoindentation in the human femur. J Biomech 32(10):1005–1012

Chapter 3
Mechanics of Biofluids
and Computational Analysis

The behavior of fluid that accounts for approximately 60% of our body weight obeys the principles of fluid mechanics. A fluid has no preferred shape. It cannot withstand any tendency by applied forces to deform it in a way which leaves the volume unchanged. It may be liquid or gas. Gases can be compressed more readily; motions appreciable pressure results in much larger changes in density than in a liquid.

In order to predict internal and external flows in the field of an interest, we often rely on numerical methods rather than analytical approaches. As the problem is more practical (i.e., three-dimensional, non-Newtonian, unsteady and interaction with solid objects), it becomes more difficult to idealize situations such that analytical approaches can be used. The widespread availability of powerful computers together with efficient solution algorithms and sophisticated pre- and post-processing facilities enable the use of computational fluid dynamics (CFD). In fact, research in the field of biofluid has been highly progressed since the 1990s when computers become available at a low cost. Many investigators including us started from fairly simple objects to quite complicated system and widened the horizon of the computational research not only for fundamental understandings but also for direct or indirect clinical applications.

While the users who started their research career from 1990s had a long learning curve with developments of CFD and are aware of limitations and problems in applying CFD to biofluid problems, new users who just abruptly dived into the highly-developed world of CFD do not always get the sufficient amount of time to obtain the fundamentals of fluid dynamics and of the numerical skills used in CFD. Thus they often face difficulties in applying CFD to biofluid problems. This chapter intends to help those readers understand fundamentals of fluid mechanics and the theoretical background required for the effective use of CFD in biofluid problems. For more advanced readers, this chapter describes issues and problems in dealing with biofluid problems in silico.

Keywords Biofluid mechanics • Dimensionless number • Discretization • Equation of continuity • Finite volume method • Fluid–structure interaction • Navier–Stokes equation • Particle method

M. Tanaka et al., *Computational Biomechanics*, A First Course in "In Silico Medicine" 3, 87
DOI 10.1007/978-4-431-54073-1_3, © Springer 2012

3.1 Fundamentals of Fluid Mechanics

3.1.1 Viscous and Inviscid Fluids

Viscosity is one of the most important natures of fluid. Viscosity is a measure of the resistance of a fluid deformed by shear stress. The unit of viscosity in SI unit is [Pa s]. Fluid can be classified in terms of the presence of viscosity. A fluid having viscosity is called viscous fluid, whereas that having no viscosity is called inviscid fluid. In reality, all kinds of fluid have viscosity except liquefied helium at extremely low temperature (<2.17 K), a phenomenon called superfluidity. In this sense, all kinds of fluid are viscous fluid. However, in practice, a fluid is treated as inviscid when viscous effects are negligibly small.

3.1.2 Newtonian and Non-Newtonian Fluids

Viscous fluid is further classified based on whether a Newton's law of viscosity works or not. A fluid that obeys the Newton's law of viscosity (i.e. shear stress increases linearly with strain rate) is called a Newtonian fluid. Air, water and some of oils are a Newtonian fluid. In contrast, a fluid whose shear-strain rate relationship is not described with the Newton's law of viscosity is called a non-Newtonian fluid. Strictly speaking, biological fluids such as blood and joint fluid belong to non-Newtonian fluids, although they are treated as Newtonian fluids in various studies in practice.

3.1.3 Compressible and Incompressible Fluids

Fluid can also be classified in terms of its compressibility. A fluid whose density varies significantly in response to a change in pressure is called compressible fluid, whereas a fluid whose material density is always constant is called incompressible fluid. In fact, all fluids (even solids) are compressible. If a fluid were incompressible, sound speed would be infinite, which contradicts the Einstein's theory of special relativity that describes that any signals do not travel faster than light. Therefore, incompressibility is a sort of approximation for describing fluid nature. An arising question is then when the approximation of incompressibility can be made. In general, compressibility of a fluid is speculated with the Mach number, M;

$$M = \frac{U}{c} \tag{3.1}$$

where U is a characteristic velocity of flow and c is the local sound speed. When the Mach number M is smaller than 0.3, compressibility of a fluid is typically considered insignificant. In this case, the fluid is usually treated as an incompressible fluid. In contrast, for a fluid having the Mach number M, larger than 0.3, compressible effects should be taken into account. When the characteristic velocity of flow exceeds the local sound speed, compressible effects became apparent, and shock wave is generated. In biological fluid, the Mach number is usually much smaller than 0.1, thus, a fluid can be treated as incompressible.

3.2 Dimensionless Numbers

Dimensionless numbers are useful in characterizing mechanical behavior of flow because they provide us insights into the relative importance of competing forces involved in a phenomenon of an interest. They also help us rescale a phenomenon. Some important dimensionless numbers in biofluid mechanics are Reynolds number (Re) and Womersley number (α). Those give flow characteristics, to say, laminar or turbulent, and the degree of flow oscillation.

3.2.1 Reynolds Number

When flow is driven through a channel at a steady state, flow is characterized by the inertia effect for dynamic pressure and viscous effect for shearing stress. The Reynolds number represents the ratio of these two factors. It is defined by

$$\text{Re} = \frac{Intertia\ force}{Viscous\ force} = \frac{\rho UL}{\mu} \tag{3.2}$$

where U is a characteristic velocity, L is a characteristic length, ρ is the density of the fluid and μ is its dynamic viscosity. Definition of the "characteristic" velocity and length are different for different problems. For a flow passing though a cylindrical flow channel, U is a spatially mean flow velocity over the cross-section of channel, and L is the channel diameter. For more complex problems, the way of defining characteristic velocity and length is not easy. Sometimes, more than two Reynolds numbers are required to represent the flow.

"Inertia" is the resistance of any physical object to change its motion or velocity. If an object has a large inertia, it will show a large resistance against a change in its velocity. On the other hand, an object with a small inertia will start a new motion or stop instantaneously upon the application of external or internally generated force. "Viscosity" is the resistance of a fluid which is being deformed by shear stresses. The viscosity of a fluid serves drag on an object that moves in the fluid. For such an object, inertia hence strives to keep the object going, whereas viscosity tries to stop it.

Fig. 3.1 (**a**) Experimental
setup for observing laminar
and turbulent flows. (**b**)
Laminar flow. (**c**) Turbulent
flow

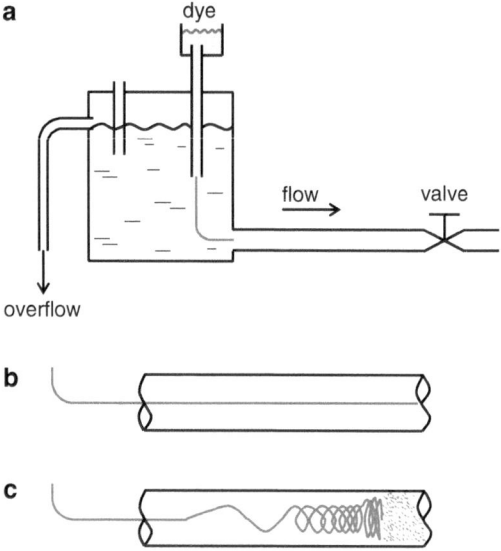

Early experiments with flow in a cylindrical flow channel by Osborne Reynolds demonstrated two representative patterns of flow, laminar and turbulent. His experimental setup is illustrated in Fig. 3.1. A transparent cylinder is attached to a constant-head tank filled with water that gives a constant preload. Dye is injected delicately into the cylindrical channel from its inlet, and the resulting flow pattern is observed. For low flow rates, the dye pattern forms a single line as illustrated in Fig. 3.1b. There is no lateral mixing of fluid, and fluid flows with keeping layers which never overlap each other. In other words, a fluid particle in one layer always stays in that layer. This type of flow is called laminar flow. As the flow rate is increased and exceeds a certain point, flow starts to disperse, showing mixing of dyes as in Fig. 3.1c. This type of flow is called turbulent flow. When flow is turbulent, unsteady vortices appear on many scales and interact with each other. The vortices mix the fluid by moving particles tortuously about the cross-section.

The experiment shown in Fig. 3.1 was repeated with various fluid properties, different channel diameters and flow rates. The results with a dimensional analysis revealed that the criterion between laminar and turbulent flows is determined with the Reynolds number. The inertia of fluid flows is caused by non-linear interactions within the flow field. The non-linearities may cause instabilities in the flow to grow, and therefore the flow can get turbulent when inertial effects are dominant, that is, for large Reynolds numbers. For small Reynolds number, on the other hand, the flow always is laminar. The Reynolds number, above which flow may be turbulent, is called the critical Reynolds number. For a cylindrical flow, it is about 2,300. Note that the flow with the Reynolds number larger than the critical Reynolds number is not always turbulent. The transition from laminar flow to turbulent flow depends on mechanical vibration, surface roughness and intrinsic

Fig. 3.2 Schematic drawing of a difference in flow patterns between small and large

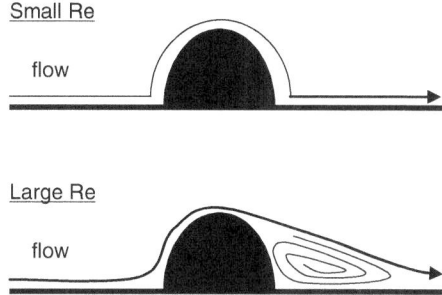

disturbances in the flow. Some experiments demonstrated a laminar flow up to the Reynolds number of 20,000. As illustrated in Fig. 3.2, flow with a small Reynolds number goes along the geometry of a flow channel. However, with an increase in the Reynolds number, flow separation may occur and vortices are created behind an obstacle.

Exercise 3.1 Calculate the Reynolds number when a fluid with the density of 1,000 kg/m^3 and the viscosity of 0.004 Pa s flows at a spatially mean velocity of 1 m/s through a cylindrical channel with the diameter of 0.2 m.

3.2.2   Womersley Number

Womersley number is a dimensionless parameter to express pulsatility of oscillating flow. It is expressed as a ratio of the oscillatory inertia to viscous momentum. The Womersley number, usually denoted α, is defined by

$$\alpha = \frac{\text{oscillartory inertia}}{\text{viscous momentum}} = \sqrt{\frac{\rho U \cdot \omega L}{\mu\left(\frac{U}{L^2}\right)}} = L\left(\frac{\omega}{\nu}\right)^{\frac{1}{2}} \tag{3.3}$$

where L is a characteristic length, ω is the angular frequency of oscillations, and ν, ρ, μ are a kinematic viscosity, density and dynamic viscosity of the fluid respectively.

Where does the Womersley number come from? Although it is slightly beyond the scope of this section, we briefly explain the origin of the Womersley number. Suppose that a pulsatile flow of an incompressible viscous fluid in a cylindrical channel is driven by an oscillating pressure gradient across an axial direction (x-axis) of the channel. If the channel is rigid, the motion of fluid is expressed as

$$\frac{\partial u}{\partial t} = -\frac{1}{\rho}\frac{\partial p}{\partial x} + \frac{\mu}{\rho}\left(\frac{\partial^2 u}{\partial r^2} + \frac{1}{r}\frac{\partial u}{\partial r}\right) \tag{3.4}$$

where u is an axial velocity that is a function of radial coordinate r and time t, p is pressure, r is a radial coordinate, R is the radius of the channel, ρ is the density of a fluid and μ is the dynamic viscosity. Assuming the pressure gradient is given by

$$\frac{\partial p}{\partial x} = Ae^{i\omega t} \tag{3.5}$$

where ω is angular velocity of the oscillating pressure and i is a complex number, and letting

$$u(t) = u\,e^{i\omega t}, \tag{3.6}$$

by substitution of (3.4) and (3.5) into (3.6), we get

$$\frac{d^2 u}{dr^2} + \frac{1}{r}\frac{du}{dr} - \frac{i\omega\rho}{\mu}u = \frac{A}{\mu} \tag{3.7}$$

The boundary conditions here are

$$\begin{cases} u = 0 & at\ r = R \\ \dfrac{du}{dr} = 0 & at\ r = 0 \end{cases} . \tag{3.8}$$

The solution of (3.7) under these boundary conditions are written as

$$u(r,t) = \frac{A}{i\omega\rho}\left[1 - \frac{J_0\left(i^{3/2}r\sqrt{\omega/v}\right)}{J_0\left(i^{3/2}R\sqrt{\omega/v}\right)}\right]e^{i\omega t} \tag{3.9}$$

where J_0 is a Bessel function of the first kind of order zero, and quantity $R\sqrt{\omega/v}$ is the Womersley number that is introduced at the top of this section.

How does a flow behavior vary with a change in the Womersley number? Figure 3.3 shows velocity profiles of a flow in a channel with the diameter of 0.02 m. Here the density of fluid is 1,000 kg/m^3, and the pressure gradient with the maximum of 30 Pa is given sinusoidally at a frequency of 10 Hz. In this case, a flow is driven by a pressure gradient in a flow direction that changes with time. When the Womersley number is sufficiently low ($\alpha < 1$), nearly no phase difference between pressure and flow exists. The velocity profile remains parabolic at all times. With an increase in the frequency, a central part of the velocity profile starts to delay behind changes in the pressure gradient. In contrast, flow near the wall does not lag behind and follows changes in the pressure gradient. A further increase in the Womersley number ($\alpha = 12$) resulted in a drastic change in the velocity profile. The velocity delays for a phase of $\pi/2$ from the pressure, and it takes a rather flat profile.

Exercise 3.2 Calculate the Womersley number when a fluid with the density of 1,000 kg/m^3 and the viscosity of 0.004 Pa s oscillates at the frequency of 2 Hz in a cylindrical channel with the diameter of 0.2 m.

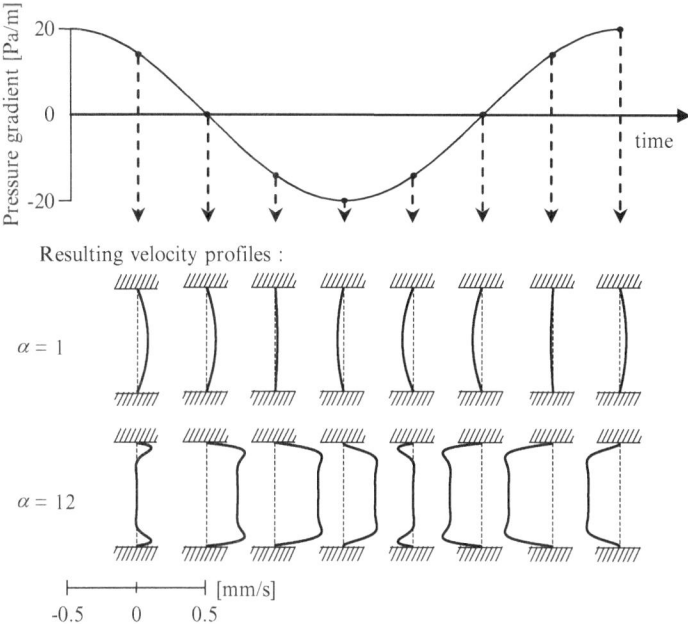

Fig. 3.3   Velocity profiles of a flow between the parallel plates during oscillating pressure gradient at the Womersley number of 1 and 10

3.3   Eulerian and Lagrangian Representations of Fluid Flow

There are two ways in describing flows, Lagrangian and Eulerian descriptions. The Lagrangian description often appears in solid mechanics and involves describing the motion of flow with movements of particles. This is similar to describing the motion of a mass point falling under the influence of gravity. At anytime, its position is described as a distance from its initial position where the mass point starts to fall. The Eulerian description is more used in fluid mechanics. In this description, the motion of flow is expressed with fluid mechanical quantities such as velocity and pressure at fixed points of a reference frame in space as time varies. In other words, this is a field description. Because a fluid is a continuous medium— a fluid volume contiguously changes its shape and different particles within the fluid volume travel at different speeds, the Lagrangian description is harder to apply though easier to understand than the Eulerian description.

In the Lagrangian description, fluid is regarded as collective mass of particles. The motion of each particle is described by its position in a global coordinate system as a function of time. Suppose a particle at $\mathbf{r_0} = (x_0, y_0, z_0)$ at time $t = t_0$ moves to $\mathbf{r} = (x, y, z)$ at time $t = t_0 + \Delta t$. As the position of each particle is dependent on its initial position and time, the position $\mathbf{r} = (x, y, z)$ is described as a function of $\mathbf{r_0} = (x_0, y_0, z_0)$ and t;

$$\mathbf{r} = \mathbf{r}(x_0, y_0, z_0, t) = \mathbf{r}(\mathbf{r_0}, t). \tag{3.10}$$

As in this equation, in the Lagrangian description, each particle is identified with its reference position at time t and its movement is traced with those. The velocity $\mathbf{u} = (u, v, w)$ and acceleration $\mathbf{a}$ of a particle from $t = t_0$ to $t = t_0 + \Delta t$ are given by

$$\mathbf{u} = \lim_{\Delta t \to 0} \frac{\mathbf{r}(\mathbf{x_0}, t + \Delta t) - \mathbf{r}(\mathbf{x_0}, t)}{\Delta t} = \frac{\partial \mathbf{r}}{\partial t}, \tag{3.11}$$

$$\mathbf{a} = \frac{\partial^2 \mathbf{r}}{\partial t^2}. \tag{3.12}$$

In contrast, in the Eulerian description, flow at any instant is expressed as a function of fixed spatial coordinates and time. In this framework, the velocity of a fluid particle, (u, v, w) at (x, y, z) is represented by

$$\mathbf{u} = \mathbf{u}(x, y, z, t). \tag{3.13}$$

Now, readers should pay attentions to (x, y, z) in (3.10) and (3.13). In the Lagrangian description, (x, y, z) are dependent variables, whereas they are independent variables in the Eulerian description. Suppose that a fluid particle at (x, y, z) that has a velocity of (u, v, w) moves to $(x + \Delta x, y + \Delta y, z + \Delta z)$ and its velocity changed to $(u + \Delta u, v + \Delta v, w + \Delta w)$. An increment in the x-component of velocity, Δu, is

$$\Delta u = u + \Delta u - u$$
$$= u(x + \Delta x, y + \Delta y, z + \Delta z, t + \Delta t) - u(x, y, z, t). \tag{3.14}$$

Using the Taylor expansion, we rewrite $u(x + \Delta x, y + \Delta y, z + \Delta z, t + \Delta t)$ as

$$u(x + \Delta x, y + \Delta y, z + \Delta z, t + \Delta t) =$$
$$u(x, y, z, t) + \frac{\partial u}{\partial x}\Delta x + \frac{\partial u}{\partial y}\Delta y + \frac{\partial u}{\partial z}\Delta z + \frac{\partial u}{\partial t}\Delta t + O(\Delta x^2, \Delta y^2, \Delta z^2, \Delta t^2) \tag{3.15}$$

where $O(\Delta x^2, \Delta y^2, \Delta z^2, \Delta t^2)$ is the remainder of the Taylor series, which is the terms having $\Delta x^2, \Delta y^2, \Delta z^2, \Delta t^2$ or larger orders of them. Substituting (3.15) into (3.14), we obtain

$$\Delta u = \frac{\partial u}{\partial x}\Delta x + \frac{\partial u}{\partial y}\Delta y + \frac{\partial u}{\partial z}\Delta z + \frac{\partial u}{\partial t}\Delta t + O(\Delta x^2, \Delta y^2, \Delta z^2, \Delta t^2). \tag{3.16}$$

Because a distance is a product of velocity and time like $\Delta x = u\Delta t$, $\Delta y = v\Delta t$ and $\Delta z = w\Delta t$, (3.16) becomes

$$\Delta u = \frac{\partial u}{\partial x}u\Delta t + \frac{\partial u}{\partial y}v\Delta t + \frac{\partial u}{\partial z}w\Delta t + \frac{\partial u}{\partial t}\Delta t$$
$$+ O((u\Delta t)^2, (v\Delta t)^2, (w\Delta t)^2, \Delta t^2). \tag{3.17}$$

The acceleration of the particle in the x-direction at (x, y, z) can be calculated by taking a limit of $\Delta u/\Delta t$ with Δt to zero;

$$\lim_{\Delta t \to 0}\frac{\Delta u}{\Delta t} = \frac{\partial u}{\partial x}u + \frac{\partial u}{\partial y}v + \frac{\partial u}{\partial z}w + \frac{\partial u}{\partial t} + O(u^2\Delta t, v^2\Delta t, w^2\Delta t, \Delta t)$$
$$= \frac{\partial u}{\partial x}u + \frac{\partial u}{\partial y}v + \frac{\partial u}{\partial z}w + \frac{\partial u}{\partial t}. \tag{3.18}$$

Similarly, the acceleration in the y and z directions are written as

$$\lim_{\Delta t \to 0}\frac{\Delta v}{\Delta t} = \frac{\partial v}{\partial x}u + \frac{\partial v}{\partial y}v + \frac{\partial v}{\partial z}w + \frac{\partial v}{\partial t}, \tag{3.19a}$$

$$\lim_{\Delta t \to 0}\frac{\Delta w}{\Delta t} = \frac{\partial w}{\partial x}u + \frac{\partial w}{\partial y}v + \frac{\partial w}{\partial z}w + \frac{\partial w}{\partial t}. \tag{3.19b}$$

In a vector form, the acceleration at (x, y, z) in the Eulerian description is given by

$$\frac{D\mathbf{u}}{Dt} = \frac{\partial \mathbf{u}}{\partial t} + (\mathbf{u} \cdot \nabla)\mathbf{u} \tag{3.20}$$

where ∇ is a derivative operator $\nabla = (\partial/\partial x, \partial/\partial y, \partial/\partial z)$. D/Dt is called material derivative. The first term of (3.20) represents the acceleration due to a temporal change in a flow field, and the second term does the acceleration due to convection. If a flow is steady, the first term vanishes.

Exercise 3.3 Derive the material derivative of velocities in the y and z directions, Dv/Dt and Dz/Dt.

3.4   Governing Equation of Fluid Flow

3.4.1   *Equation of Continuity*

The equation of continuity is a law of conservation of mass in fluid mechanics. When deriving the equation of continuity, the use of the Eulerian description is easier than the Lagrangian description, because it is difficult to track a particular volume of fluid that always moves and deforms with flowing.

Fig. 3.4 Control volume
for deriving the equation
of continuity

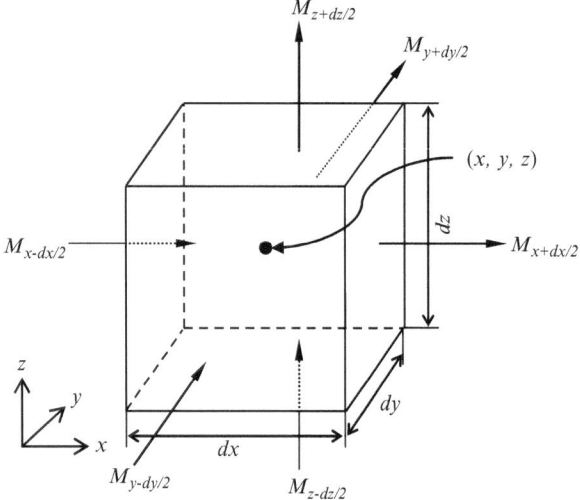

Suppose a cubic control volume in a flow field as shown in Fig. 3.4. The center of this control volume is (x, y, z). A mass of the flow coming in this control volume through the face at $x-dx/2$ is

$$M_{x-dx/2} = (\rho u A)_{x-dx/2} \tag{3.21}$$

where ρ is density, u is a velocity in x-direction and A is an area of the face of the control volume at $x-dx/2$. The area of the face at $x-dx/2$ is simply

$$A_{x-dx/2} = dydz. \tag{3.22}$$

On the other hand, a mass of the flow that goes out from the face at $x + dx/2$ is

$$M_{x+dx/2} = (\rho u)_{x+dx/2} dydz. \tag{3.23}$$

In a similar fashion, a mass of the flow passing through other faces is written as

$$M_{y-dy/2} = (\rho v)_{y-dy/2} dxdz \tag{3.24a}$$

$$M_{y+dy/2} = (\rho v)_{y+dy/2} dxdz \tag{3.24b}$$

$$M_{z-dz/2} = (\rho w)_{z-dz/2} dxdy \tag{3.24c}$$

$$M_{z+dz/2} = (\rho w)_{z+dz/2} dxdy \tag{3.24d}$$

where v and w are velocities in the y and z directions, respectively. Because a mass of the control volume is $\rho dxdydz$, a change rate of the mass of the control volume is given by

$$\frac{\partial \rho}{\partial t} dxdydz \tag{3.25}$$

which must be equal to the sum of mass fluxes from all faces of the control volume;

$$\frac{\partial \rho}{\partial t} dxdydz = M_{x-dx/2} - M_{x+dx/2} + M_{y-dy/2} - M_{y+dy/2} + M_{z-dz/2} - M_{z+dz/2}$$

$$= (\rho u)_{x-dx/2} dydz - (\rho u)_{x+dx/2} dydz$$

$$+ (\rho v)_{y-dy/2} dxdz - (\rho v)_{y+dy/2} dxdz$$

$$+ (\rho w)_{z-dz/2} dxdy - (\rho w)_{z+dz/2} dxdy$$

$$(3.26)$$

Using the Taylor expansion, we rewrite each term on the right hand side in the above equation. For instance,

$$(\rho u)_{x-dx/2} dydz = (\rho u) dydz - \frac{\partial (\rho u)}{\partial x} \frac{dx}{2} dydz + \frac{1}{2!} \frac{\partial^2 (\rho u)}{\partial x^2} \left(\frac{dx}{2}\right)^2 dydz - \cdots, \quad (3.27a)$$

$$(\rho u)_{x+dx/2} dydz = (\rho u) dydz + \frac{\partial (\rho u)}{\partial x} \frac{dx}{2} dydz + \frac{1}{2!} \frac{\partial^2 (\rho u)}{\partial x^2} \left(\frac{dx}{2}\right)^2 dydz + \cdots. \quad (3.27b)$$

Substituting those in (3.26), we obtain

$$\frac{\partial \rho}{\partial t} dxdydz = -\frac{\partial (\rho u)}{\partial x} dxdydz - \frac{\partial (\rho v)}{\partial y} dxdydz - \frac{\partial (\rho w)}{\partial z} dxdydz$$

$$+ O(dx^2, dy^2, dz^2) \quad (3.28)$$

where $O(dx^2, dy^2, dz^2)$ is all the rest of terms generated by the Taylor expansion which have dx^2, dy^2 or dz^2 or more order of dx, dy or dz. Dividing (3.28) with the volume $dxdydz$ and taking a limit of dx, dy and dz to zero, we gain

$$\frac{\partial \rho}{\partial t} = -\frac{\partial (\rho u)}{\partial x} - \frac{\partial (\rho v)}{\partial y} - \frac{\partial (\rho w)}{\partial z}. \quad (3.29)$$

Moving all terms in the right hand side to the left, we have

$$\frac{\partial \rho}{\partial t} + \frac{\partial (\rho u)}{\partial x} + \frac{\partial (\rho v)}{\partial y} + \frac{\partial (\rho w)}{\partial z} = 0. \quad (3.30)$$

Equation (3.30) is called the equation of continuity that describes the conservation of mass in fluid mechanics. In vector form, (3.30) is rewritten as

$$\frac{\partial \rho}{\partial t} + \nabla \cdot (\rho \mathbf{u}) = 0. \quad (3.31)$$

If a flow is steady (a velocity field does not change with time), the first term of (3.31) is zero. Thus, the equation of continuity for a steady flow is

$$\nabla \cdot (\rho \mathbf{u}) = 0. \quad (3.32)$$

Furthermore, if a fluid can be assumed to be incompressible, the density is constant in time and space. Thus, for an incompressible fluid, the equation of continuity is reduced to

$$\frac{\partial u}{\partial x} + \frac{\partial v}{\partial y} + \frac{\partial w}{\partial z} = 0 \tag{3.33}$$

or

$$\nabla \cdot \mathbf{u} = 0. \tag{3.34}$$

3.4.2 Navier–Stokes Equation

The Navier–Stokes equation is a law of the conservation of momentum in fluid mechanics. This equation was obtained independently by two scientists in nineteenth century, Claude Louis Marie Henri Navier and George Gabriel Stokes. As was in the equation of continuity, we use the Eulerian description in order to derive the Navier–Stokes equation. Consider the control volume of Fig. 3.5. Here we consider surface forces such as viscous and pressure forces as well as body forces including gravitational, centrifugal, Coriolis and electromagnetic forces in order to encompass various flow problems. The state of stresses acting on a fluid element is illustrated in Fig. 3.6. The suffix notation of a stress τ_{ij} indicates that the stress acts in the j-axis on a surface normal to the i-axis.

Let us start with deviating the Navier–Stokes equation in the x-direction. Figure 3.6 illustrates surface forces working in the x-direction. A net force acting on the pair of faces at $x - dx/2$ and $x + dx/2$ is

$$(p - \tau_{xx})_{x-dx/2} dydz - (p - \tau_{xx})_{x+dx/2} dydz. \tag{3.35}$$

Readers should pay attentions to positive and negative signs in (3.35) by referring to a direction of arrows in Fig. 3.6. Applying the Taylor expansion to each term, we obtain, for example,

$$p_{x-dx/2} = p - \frac{\partial p}{\partial x}\frac{dx}{2} + \frac{1}{2!}\frac{\partial p}{\partial x}\left(\frac{dx}{2}\right)^2 - \cdots, \tag{3.36a}$$

$$p_{x+dx/2} = p + \frac{\partial p}{\partial x}\frac{dx}{2} + \frac{1}{2!}\frac{\partial p}{\partial x}\left(\frac{dx}{2}\right)^2 + \cdots. \tag{3.36b}$$

Fig. 3.5 Stress components acting on the faces of a control volume

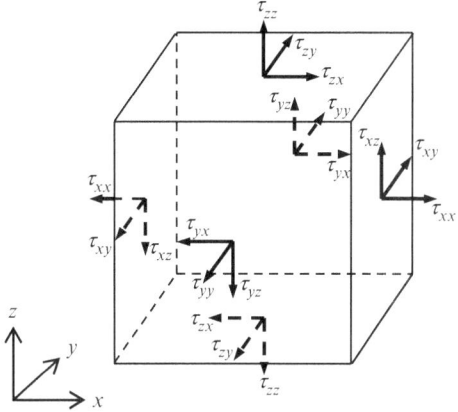

Fig. 3.6 Stress components in the x-direction which act on faces of the control volume

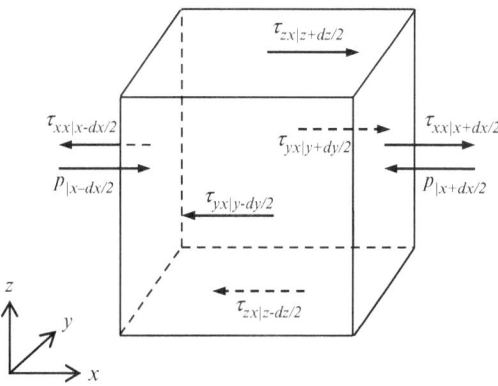

Thus, the net force in the x-direction on the pair of faces at $x-dx/2$ and $x + dx/2$ is

$$(p - \tau_{xx})_{x-dx/2}dydz - (p - \tau_{xx})_{x+dx/2}dydz = \left(-\frac{\partial p}{\partial x} + \frac{\partial \tau_{xx}}{\partial x}\right)dxdydz + O(dx^3)$$

(3.37)

where $O(dx^3)$ is the rest of terms having the order of dx^3 or more which result from the Taylor expansion. Similarly, a net force in the y-direction is

$$- \tau_{yx|y-dy/2}dxdz + \tau_{yx|y+dy/2}dxdz = \left(\frac{\partial \tau_{yx}}{\partial y}\right)dxdydz + O(dy^3)$$

(3.38)

and that in the z-direction is

$$- \tau_{zx|z-dz/2}dxdy + \tau_{zx|z+dz/2}dxdy = \left(\frac{\partial \tau_{zx}}{\partial z}\right)dxdydz + O(dz^3).$$

(3.39)

Note that the pressure term does not appear in (3.38) and (3.39). We do not have to describe a body force in details. Instead, we use f_x denote a body force per unit volume. Now we have all components of a force in the x-direction. According to the Newton's second law, the net force is equal to a product of a mass of the control volume $\rho dxdydz$ and acceleration Du/Dt. Thus, we obtain

$$\rho dxdydz \frac{Du}{Dt} = \left\{ \left(-\frac{\partial p}{\partial x} + \frac{\partial \tau_{xx}}{\partial x} \right) + \left(\frac{\partial \tau_{yx}}{\partial y} \right) + \left(\frac{\partial \tau_{zx}}{\partial z} \right) \right\} dxdydz$$
$$+ O\left(dx^3, dy^3, dz^3 \right) + f_x dxdydz. \tag{3.40}$$

Dividing (3.40) with the volume $dxdydz$ and taking a limit of dx, dy and dz to zero, we have

$$\rho \frac{Du}{Dt} = -\frac{\partial p}{\partial x} + \frac{\partial \tau_{xx}}{\partial x} + \frac{\partial \tau_{yx}}{\partial y} + \frac{\partial \tau_{zx}}{\partial z} + f_x. \tag{3.41}$$

This is the Navier–Stokes equation in the x-direction. Similarly, we gain the Navier–Stokes equation for the y and z directions as

$$\rho \frac{Dv}{Dt} = -\frac{\partial p}{\partial y} + \frac{\partial \tau_{xy}}{\partial x} + \frac{\partial \tau_{yy}}{\partial y} + \frac{\partial \tau_{zy}}{\partial z} + f_y, \tag{3.42a}$$

$$\rho \frac{Dw}{Dt} = -\frac{\partial p}{\partial z} + \frac{\partial \tau_{xz}}{\partial x} + \frac{\partial \tau_{yz}}{\partial y} + \frac{\partial \tau_{zz}}{\partial z} + f_z. \tag{3.42b}$$

Although we now have the Navier–Stokes equations, they still contain the viscous stress components as unknowns. As described in Sect. 1.2.2, viscous stresses (or stress tensors) are expressed as a function of strain rate (local deformation of a fluid element).

The Newton's law of viscosity in a three dimensional space is

$$\tau_{ij} = C_{ijkl} s_{kl} \tag{3.43}$$

where τ_{ij} is a stress tensor, C_{ijkl} is a viscosity tensor and s_{kl} is a strain rate tensor. Here the Einstein summation convention is used to write the equation. In this convention, i and j can be 1, 2 and 3 corresponding to x, y and z, and we sum over all of its possible values if an index variable appears repeatedly in a single term. An example of the Einstein summation convention is found in Footnote 3, Chapter 2. In order to model viscous forces, let us consider an isotropic fluid that has uniformity in its nature in all orientations. In fact, some of liquids, for instance blood which contain significant amount of particles (red blood cells) may exhibit anisotropic or directionality in viscous stresses as a result of deformation of particles and their aggregations. However, most of gases and liquids are isotropic and therefore the assumption of isotropy is quite useful to gain practically reasonable results in various flow problems. For isotropic materials, the viscosity tensor is given by

$$C_{ijkl} = \lambda \delta_{ij} \delta_{kl} + \mu \left(\delta_{ik} \delta_{jl} + \delta_{il} \delta_{jk} \right) \tag{3.44}$$

where δ_{ij} is the Knonecker delta, μ is the dynamic viscosity and λ is the second coefficient of viscosity. The Knonecker delta gives 1 if $i = j$, and zero unless otherwise (cf. footnote 5 in Sect. 2.1.3). Substituting (3.44) into (3.43), we gain

$$\begin{aligned}
\tau_{ij} &= \left\{ \lambda \delta_{ij} \delta_{kl} + \mu \left(\delta_{ik} \delta_{jl} + \delta_{il} \delta_{jk} \right) \right\} s_{kl} \\
&= \lambda \delta_{ij} \delta_{kl} s_{kl} + \mu \left(\delta_{ik} \delta_{jl} + \delta_{il} \delta_{jk} \right) s_{kl} \\
&= \lambda \delta_{ij} s_{kk} + \mu \left(s_{ij} + s_{ji} \right).
\end{aligned} \tag{3.45}$$

Because the strain rate tensor s_{ij} is

$$s_{ij} = \frac{1}{2} \left(\frac{\partial u_i}{\partial x_j} + \frac{\partial u_j}{\partial x_i} \right), \tag{3.46}$$

Equation (3.45) becomes

$$\tau_{ij} = \lambda \delta_{ij} \frac{\partial u_k}{\partial x_k} + \mu \left(\frac{\partial u_i}{\partial x_j} + \frac{\partial u_j}{\partial x_i} \right) \tag{3.47}$$

where δ_{ij} is the Knonecker delta, μ is the dynamic viscosity and λ is the second coefficient of viscosity. Note that λ is analogous to the first Lame's constant in solid mechanics. Let us give some examples on what (3.47) indicates. If $i = 1$ and $j = 1$, the above equation is

$$\tau_{11} = \lambda \delta_{11} \left(\frac{\partial u_1}{\partial x_1} + \frac{\partial u_2}{\partial x_2} + \frac{\partial u_3}{\partial x_3} \right) + \mu \left(\frac{\partial u_1}{\partial x_1} + \frac{\partial u_1}{\partial x_1} \right) \tag{3.48}$$

which in turn becomes

$$\tau_{xx} = \lambda \left(\frac{\partial u}{\partial x} + \frac{\partial v}{\partial y} + \frac{\partial w}{\partial z} \right) + 2\mu \frac{\partial u}{\partial x}. \tag{3.49}$$

One more example with $i = 1$ and $j = 2$ is

$$\tau_{12} = \lambda \delta_{12} \left(\frac{\partial u_1}{\partial x_1} + \frac{\partial u_2}{\partial x_2} + \frac{\partial u_3}{\partial x_3} \right) + \mu \left(\frac{\partial u_1}{\partial x_2} + \frac{\partial u_2}{\partial x_1} \right) \tag{3.50}$$

which is reduced to

$$\tau_{xy} = \mu \left(\frac{\partial u}{\partial y} + \frac{\partial v}{\partial x} \right). \tag{3.51}$$

Now, we substitute these stress tensors into (3.41) and rewrite the Navier–Stokes equation in the x-direction as

$$\rho \frac{Du}{Dt} = -\frac{\partial p}{\partial x} + \frac{\partial}{\partial x}\left(\lambda\left(\frac{\partial u}{\partial x} + \frac{\partial v}{\partial y} + \frac{\partial w}{\partial z}\right) + 2\mu\frac{\partial u}{\partial x}\right)$$
$$+ \frac{\partial}{\partial y}\left(\mu\left(\frac{\partial u}{\partial y} + \frac{\partial v}{\partial x}\right)\right) + \frac{\partial}{\partial z}\left(\mu\left(\frac{\partial u}{\partial z} + \frac{\partial w}{\partial x}\right)\right) + f_x. \qquad (3.52)$$

Some mathematical rearrangements yield

$$\rho \frac{Du}{Dt} = -\frac{\partial p}{\partial x} + \frac{\partial}{\partial x}\left(\lambda\left(\frac{\partial u}{\partial x} + \frac{\partial v}{\partial y} + \frac{\partial w}{\partial z}\right)\right)$$
$$+ \left[\frac{\partial}{\partial x}\left(\mu\frac{\partial u}{\partial x}\right) + \frac{\partial}{\partial y}\left(\mu\frac{\partial u}{\partial y}\right) + \frac{\partial}{\partial z}\left(\mu\frac{\partial u}{\partial z}\right)\right]$$
$$+ \frac{\partial}{\partial x}\left(\mu\left(\frac{\partial u}{\partial x} + \frac{\partial v}{\partial y} + \frac{\partial w}{\partial z}\right)\right) + f_x. \qquad (3.53)$$

It is not too difficult to verify the Navier–Stokes equations for the y and z directions given as

$$\rho \frac{Dv}{Dt} = -\frac{\partial p}{\partial y} + \frac{\partial}{\partial y}\left(\lambda\left(\frac{\partial u}{\partial x} + \frac{\partial v}{\partial y} + \frac{\partial w}{\partial z}\right)\right)$$
$$+ \left[\frac{\partial}{\partial x}\left(\mu\frac{\partial v}{\partial x}\right) + \frac{\partial}{\partial y}\left(\mu\frac{\partial v}{\partial y}\right) + \frac{\partial}{\partial z}\left(\mu\frac{\partial v}{\partial z}\right)\right]$$
$$+ \frac{\partial}{\partial y}\left(\mu\left(\frac{\partial u}{\partial x} + \frac{\partial v}{\partial y} + \frac{\partial w}{\partial z}\right)\right) + f_y, \qquad (3.54)$$

$$\rho \frac{Dw}{Dt} = -\frac{\partial p}{\partial z} + \frac{\partial}{\partial z}\left(\lambda\left(\frac{\partial u}{\partial x} + \frac{\partial v}{\partial y} + \frac{\partial w}{\partial z}\right)\right)$$
$$+ \left[\frac{\partial}{\partial x}\left(\mu\frac{\partial w}{\partial x}\right) + \frac{\partial}{\partial y}\left(\mu\frac{\partial w}{\partial y}\right) + \frac{\partial}{\partial z}\left(\mu\frac{\partial w}{\partial z}\right)\right]$$
$$+ \frac{\partial}{\partial z}\left(\mu\left(\frac{\partial u}{\partial x} + \frac{\partial v}{\partial y} + \frac{\partial w}{\partial z}\right)\right) + f_z. \qquad (3.55)$$

Physics of the second coefficient of viscosity λ is not well known. Nevertheless, its effect is usually quite small in practice. For gases, good approximation can be made for $\lambda = -2/3\ \mu$. For incompressible fluids including liquids, the conservation of mass in (3.33) is always satisfied, which subsequently vanishes the term of the second coefficient of viscosity and the second viscosity term. As a consequence, the Navier–Stokes equations for an incompressible fluid is

$$\rho \frac{Du}{Dt} = -\frac{\partial p}{\partial x} + \left[\frac{\partial}{\partial x}\left(\mu\frac{\partial u}{\partial x}\right) + \frac{\partial}{\partial y}\left(\mu\frac{\partial u}{\partial y}\right) + \frac{\partial}{\partial z}\left(\mu\frac{\partial u}{\partial z}\right)\right] + f_x, \qquad (3.56a)$$

$$\rho \frac{Dv}{Dt} = -\frac{\partial p}{\partial y} + \left[\frac{\partial}{\partial x} \left(\mu \frac{\partial v}{\partial x} \right) + \frac{\partial}{\partial y} \left(\mu \frac{\partial v}{\partial y} \right) + \frac{\partial}{\partial z} \left(\mu \frac{\partial v}{\partial z} \right) \right] + f_y, \qquad (3.56b)$$

$$\rho \frac{Dw}{Dt} = -\frac{\partial p}{\partial z} + \left[\frac{\partial}{\partial x} \left(\mu \frac{\partial w}{\partial x} \right) + \frac{\partial}{\partial y} \left(\mu \frac{\partial w}{\partial y} \right) + \frac{\partial}{\partial z} \left(\mu \frac{\partial w}{\partial z} \right) \right] + f_z. \qquad (3.56c)$$

In the vector form,

$$\rho \frac{D\mathbf{u}}{Dt} = -\nabla p + \nabla \cdot (\mu \nabla \mathbf{u}) + \mathbf{f}. \qquad (3.57)$$

Furthermore, if the viscosity is constant in space (usually it is), it becomes

$$\rho \frac{D\mathbf{u}}{Dt} = -\nabla p + \mu \nabla^2 \mathbf{u} + \mathbf{f}. \qquad (3.58)$$

Exercise 3.4 Derive the Navier–Stokes equation in the y and z directions (3.54) and (3.55).

3.5 Euler-Based Computational Fluid Dynamics

3.5.1 *Discretization*

As shown in Sect. 3.4, governing equations of a fluid are written as partial differential equations. An analytical solution of partial differential equations is obtained only in special and simple cases. Thus, we have to rely on numerical means to gain solutions. Numerically solving the governing equations of fluid dynamics is called computational fluid dynamics (CFD). Discretization is the first step of CFD to approximate a derivate with discrete values for solving differential equations by numerical means. By discretization we transfer continuous functions and equations into discrete quantities.

Suppose that there are three nodes positioned equidistantly with a space of h along the x-axis as shown in Fig. 3.7.

One way of evaluating a derivate $d\phi/dx$ with quantities at nodes is

$$\frac{d\phi}{dx} \cong \frac{\phi(x+h) - \phi(x)}{h}. \qquad (3.59)$$

This is called forward difference method because the derivate $d\phi/dx$ is discretized with the physical quantity at $x + h$ and x. This approximation approaches to a true value of $d\phi/dx$ as h is ultimately small. The problem is that it is difficult to understand how erroneous the discretized value on the right hand side of (3.59)

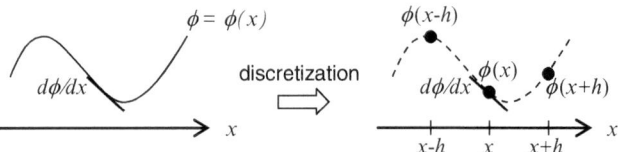

Fig. 3.7 Conceptual image of discretization

with a choice of the increment size of h is in comparison to the actual derivative value. Assuming that $\phi(x)$ is spatially continuous, we apply the Taylor expansion to $\phi(x + h)$ with respect to x,

$$\phi(x + h) = \phi(x) + \frac{d\phi}{dx}h + \frac{1}{2!}\frac{d^2\phi}{dx^2}h^2 + \frac{1}{3!}\frac{d^3\phi}{dx^3}h^3 + \cdots. \qquad (3.60)$$

With some mathematical rearrangements, we obtain

$$\frac{d\phi}{dx} = \frac{\phi(x + h) - \phi(x)}{h} - \frac{1}{2!}\frac{d^2\phi}{dx^2}h - \frac{1}{3!}\frac{d^3\phi}{dx^3}h^2 - \cdots. \qquad (3.61)$$

A comparison between (3.59) and (3.61) gives that an error ε resulting from the discretization is

$$\varepsilon = -\frac{1}{2!}\frac{d^2\phi}{dx^2}h - \frac{1}{3!}\frac{d^3\phi}{dx^3}h^2 - \cdots. \qquad (3.62)$$

This error is called truncation error, since this is truncated by discretization. Because the first term in (3.62) is the largest source of error, the accuracy of discretization is determined by the order of h in the first term of truncated terms. Thus, we call that the discretization of (3.59) is first-order accurate. There are other ways of discretization. Applying the Taylor expansion to $\phi(x\text{-}h)$ with respect to x, we have

$$\phi(x - h) = \phi(x) - \frac{d\phi}{dx}h + \frac{1}{2!}\frac{d^2\phi}{dx^2}h^2 - \frac{1}{3!}\frac{d^3\phi}{dx^3}h^3 + \cdots. \qquad (3.63)$$

Rearrangements of (3.63) as we did for (3.60) give

$$\frac{d\phi}{dx} = \frac{\phi(x) - \phi(x - h)}{h} + \frac{1}{2!}\frac{d^2\phi}{dx^2}h - \frac{1}{3!}\frac{d^3\phi}{dx^3}h^2 + \cdots. \qquad (3.64)$$

This is called backward difference method having first-order accuracy in space. Subtraction of (3.63) from (3.60) yields

$$\phi(x + h) - \phi(x - h) = 2\frac{d\phi}{dx}h + \frac{2}{3!}\frac{d^3\phi}{dx^3}h^3 + \cdots. \qquad (3.65)$$

Fig. 3.8 Wave propagation
of function ϕ at the speed of u

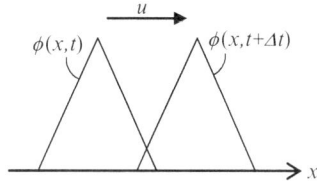

Rewriting (3.65) with $d\phi/dx$ on the left hand side, we have

$$\frac{d\phi}{dx} = \frac{\phi(x+h) - \phi(x-h)}{2h} - \frac{2}{3!}\frac{d^3\phi}{dx^3}h^2 + \cdots. \tag{3.66}$$

This is called central difference method that is second-order accurate in space, meaning that the central difference method is more accurate than the forward difference method and the back difference method. For deviating the second derivative $d^2\phi/dx^2$, we add (3.60) to (3.63);

$$\phi(x+h) + \phi(x-h) = 2\phi(x) + \frac{2}{2!}\frac{d^2\phi}{dx^2}h^2 + \frac{2}{4!}\frac{d^4\phi}{dx^4}h^4 + \cdots. \tag{3.67}$$

After some rearrangements, we obtain

$$\frac{d^2\phi}{dx^2} = \frac{\phi(x+h) - 2\phi(x) + \phi(x-h)}{2h^2} - \frac{2}{4!}\frac{d^4\phi}{dx^4}h^2 - \cdots. \tag{3.68}$$

This is one way of discretizing the second derivative with the second-order accuracy. Note that the discretization of the second derivative in (3.68) is just a repeated use of the forward discretization method of the first derivative, as it is clear from

$$\frac{d^2\phi}{dx^2} = \frac{\frac{\phi(x+h)-\phi(x)}{h} - \frac{\phi(x)-\phi(x-h)}{h}}{h} - \frac{2}{4!}\frac{d^4\phi}{dx^4}h^2 - \cdots. \tag{3.69}$$

According to the truncation errors between the forward, back and central difference methods, a reader may believe that the central difference method is the best discretization method for its accuracy. However, the central difference method is not almighty. Suppose that function ϕ travels at velocity u from left to right with no diffusion as illustrated in Fig. 3.8. This situation is expressed by a wave equation,

$$\frac{\partial\phi}{\partial t} + u\frac{\partial\phi}{\partial x} = 0. \tag{3.70}$$

Fig. 3.9 Discretization of the convective term by the central difference method

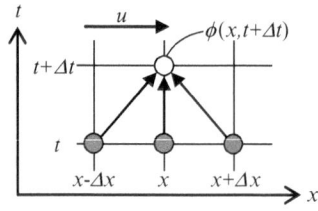

Discretization of (3.70) with the forward difference method for the temporal term and the central difference method for the convective term yields

$$\frac{\phi(x, t + \Delta t) - \phi(x, t)}{\Delta t} + u\frac{\phi(x + \Delta x, t) - \phi(x - \Delta x, t)}{2\Delta x} = 0 \qquad (3.71)$$

and thus

$$\phi(x, t + \Delta t) = \phi(x, t) + \frac{u\Delta t}{2\Delta x}\{\phi(x + \Delta x, t) - \phi(x - \Delta x, t)\}. \qquad (3.72)$$

As seen in (3.72), the discretization with the central difference method results in an equal contribution of $\phi(x + \Delta x, t)$ and $\phi(x - \Delta x, t)$ to $\phi(x, t + \Delta t)$. This is schematically shown in Fig. 3.9. Nevertheless, in the wave propagation, information is just conveyed from the upstream to the downstream. In other words, there is no propagation of information in the downstream to that in the upstream. Therefore, this discretization does not physically match with the wave propagation. In this situation, information (or physical quantity ϕ) is propagated only from the upstream and not from the downstream. In order to resolve this inadequacy, we discretize the convective term in (3.70) as

$$u\frac{\partial\phi}{\partial x} = u\frac{\phi(x, t) - \phi(x - \Delta x, t)}{\Delta x} \quad \text{if } (u > 0), \qquad (3.73a)$$

$$u\frac{\partial\phi}{\partial x} = u\frac{\phi(x + \Delta x, t) - \phi(x, t)}{\Delta x} \quad \text{if } (u < 0). \qquad (3.73b)$$

This treatment is called upwind difference method. This method takes into account the flow direction such that the direction of propagation of information in a flow field is more properly simulated. The concept of the upwind difference method is schematically shown in Fig. 3.10. The upwind difference method above is first-order accuracy, but more suitable for discretizing a convective term than the central difference method in particular when a flow is strongly convective.

For discretizing the Navier–Stokes equation in the next section, it would be helpful to show the way of evaluating a flux of ϕ at points w and e with the values at nodal points in Fig. 3.11 by the upwind difference method;

$$(\rho u\phi)_w = \phi_W\langle(\rho u)_w, 0\rangle - \phi_P\langle-(\rho u)_w, 0\rangle \qquad (3.74a)$$

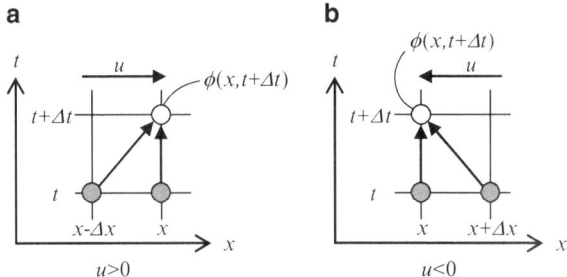

Fig. 3.10 Discretization of the convective term by the upwind difference method

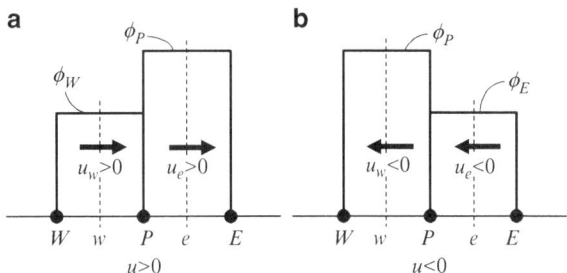

Fig. 3.11 Evaluation of the a flux of ϕ at points w and e by the upwind difference method

$$(\rho u \phi)_e = \phi_P \langle (\rho u)_e, 0 \rangle - \phi_E \langle -(\rho u)_e, 0 \rangle \tag{3.74b}$$

where $< A, B >$ means that we take a larger value between A and B.

3.5.2 Finite Volume Method

In Euler-based computer fluid dynamics, we discretize the spatial domain into small cells to form a volume mesh or grid. There are various ways to formulate the governing equations in CFD before discretization. Among those, here we introduce the finite volume method (FVM), since it is now used in many commercial software of CFD. The FVM governing equations such as the Navier–Stokes equations and the equation of continuity are written in the conservative form to guarantee conservation of in- and out-flows of momentum and mass through a small control volume, which consequently ensures global conservation. In this section, for simplicity we describe FVM formulation of the continuity and Navier–Stokes equations in 2D space for a steady, incompressible and Newtonian fluid with no external forces, which are given as

$$\frac{\partial u}{\partial x} + \frac{\partial v}{\partial y} = 0, \tag{3.75}$$

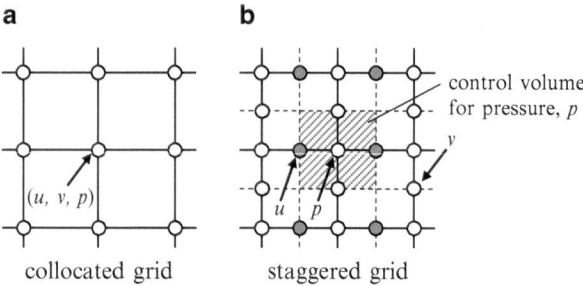

Fig. 3.12 Difference between (**a**) collocated grid and (**b**) staggered grid

$$\rho u \frac{\partial u}{\partial x} + \rho v \frac{\partial u}{\partial y} = -\frac{\partial p}{\partial x} + \mu \left(\frac{\partial^2 u}{\partial x^2} + \frac{\partial^2 u}{\partial y^2} \right), \tag{3.76a}$$

$$\rho u \frac{\partial v}{\partial x} + \rho v \frac{\partial v}{\partial y} = -\frac{\partial p}{\partial y} + \mu \left(\frac{\partial^2 v}{\partial x^2} + \frac{\partial^2 v}{\partial y^2} \right). \tag{3.76b}$$

3.5.2.1 Staggered Grid

Before formulating those governing equations by FVM, let us introduce the staggered grid. There are two types of grid used in CFD, a collocated grid and a staggered grid. On the collocated grid, the same grid is used for defining all variables. In contrast, on the staggered grid, different grids are used for different variables; scalar variables are stored in the center of a control volume, whereas vector variables such as velocity are defined at the face of the control volume. A difference between the collocated grid and the staggered grid is illustrated in Fig. 3.12.

But, why do we need the staggered grid? The importance of the staggered grid would be clear from this example. Suppose one-dimensional steady flow of an incompressible inviscid fluid. Its behavior is expressed by

$$\rho u \frac{\partial u}{\partial x} = -\frac{\partial p}{\partial x}. \tag{3.77}$$

Given the grid shown in Fig. 3.13a, the pressure term is simply discretized as

$$\begin{aligned}
\frac{\partial p}{\partial x} &= \frac{p_{i+1/2} - p_{i-1/2}}{\Delta x} \\
&= \frac{\frac{p_{i+1} + p_i}{2} - \frac{p_i + p_{i-1}}{2}}{\Delta x} \\
&= \frac{p_{i+1} - p_{i-1}}{2\Delta x}.
\end{aligned} \tag{3.78}$$

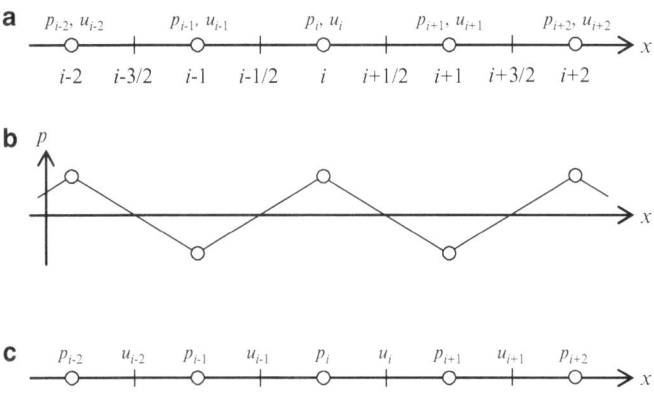

Fig. 3.13 An example for demonstrating the importance of the staggered grid. (**a**) Collocated grid, (**b**) pressure oscillation resulting from improper discretization of pressure, (**c**) staggered grid

As seen, p_i, the pressure at node i does not appear in this discretization. Suppose that velocity is zero at all nodal points. In this case, pressure p_{i+1} must be equal to p_{i-1}, but no restriction on p_i in this discretization. This means that zero value of u_i is achieved if p_{i+1} is equal to p_{i-1}, and zero value of u_{i-1} and u_{i+1} is established if p_i is equal to p_{i+2} and p_{i-2}. As a consequence, pressure may exhibit spatial oscillation as depicted in Fig. 3.13b. In this sense, pressure p_i and velocity u_i are decoupled in this grid. The remedy for this problem is to define scalar variables such as pressure at ordinary grids and vector variables at nodes staggered from the ordinary grids or on the face of a control volume such that a node where velocity is stored is sandwiched with pressure-defined nodes as in Fig. 3.13c (Harlow and Welch 1965).

3.5.2.2   Formulation of the Navier–Stokes Equation by FVM

Now we formulate and discretize the Navier–Stokes equation in the x-direction by FVM. Figure 3.14 shows a control volume and variables used for formulating velocity u.

A red region in the center represents the control volume for the Navier–Stokes equation in the x-direction. Suffixes I and J indicate nodal points where pressure and other scalar values are defined, while suffixes i and j indicate nodes where velocity u and v are defined, respectively. Because we use staggered grids, pressure p is defined at boundary faces w and e, and velocity u is defined at each corner of the control volume. We firstly integrate (3.76a) over the control volume;

$$\int_{CV}\left(\rho u\frac{\partial u}{\partial x}+\rho v\frac{\partial u}{\partial y}\right)dV = -\int_{CV}\frac{\partial p}{\partial x}dV+\int_{CV}\mu\left(\frac{\partial^2 u}{\partial x^2}+\frac{\partial^2 u}{\partial y^2}\right)dV. \qquad (3.79)$$

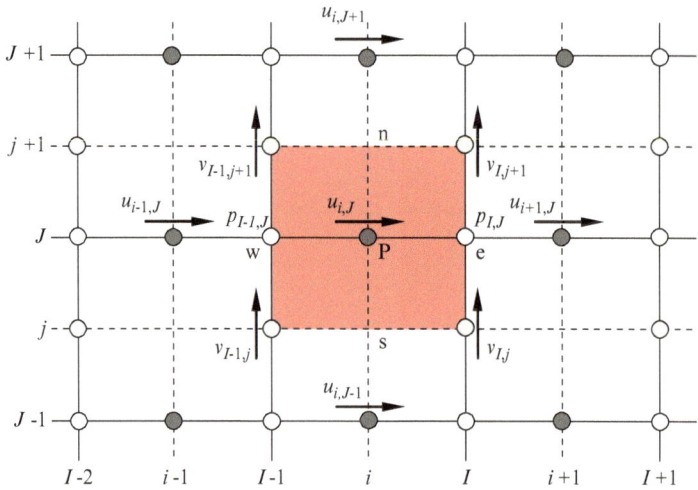

Fig. 3.14 Control volume for the Navier–Stokes equation in the x-direction

Because of complexity, a further formulation is conducted separately for each term. The first term of the left hand side is

$$\int_{CV}\left(\rho u\frac{\partial u}{\partial x}\right)dV = \iint\left(\rho u\frac{\partial u}{\partial x}\right)dxdy$$

$$= \Delta y\int_{w}^{e}\left(\rho u\frac{\partial u}{\partial x}\right)dx$$

$$= \Delta y\left\{(\rho uu)_{e} - (\rho uu)_{w}\right\} \tag{3.80}$$

where $(.)_{e}$ and $(.)_{w}$ represent the value at eastern (right) and western (left) faces of the control volume. As there are two u in each bracket, we make some tricks to linearize the equation, by which one of u is treated as a known variable and the other is left as an unknown variable as follows:

$$\int_{CV}\left(\rho u\frac{\partial u}{\partial x}\right)dV = \Delta y\left\{(\rho uu)_{e} - (\rho uu)_{w}\right\}$$

$$= \left[\Delta y\langle(\rho u)_{e},0\rangle u_{i,J} - \Delta y\langle-(\rho u)_{e},0\rangle u_{i+1,J}\right]$$

$$- \left[\Delta y\langle(\rho u)_{w},0\rangle u_{i-1,J} - \Delta y\langle-(\rho u)_{w},0\rangle u_{i,J}\right]$$

$$= \left[\Delta y\left\langle\rho\frac{u_{i+1,J}+u_{i,J}}{2},0\right\rangle u_{i,J} - \Delta y\left\langle-\rho\frac{u_{i+1,J}+u_{i,J}}{2},0\right\rangle u_{i+1,J}\right]$$

$$- \left[\Delta y\left\langle\rho\frac{u_{i,J}+u_{i-1,J}}{2},0\right\rangle u_{i-1,J} - \Delta y\left\langle-\rho\frac{u_{i,J}+u_{i-1,J}}{2},0\right\rangle u_{i,J}\right]$$

$$= \left[\langle C_{e},0\rangle u_{i,J} - \langle-C_{e},0\rangle u_{i+1,J}\right] - \left[\langle C_{w},0\rangle u_{i-1,J} - \langle-C_{w},0\rangle u_{i,J}\right]$$

$$= -\langle-C_{e},0\rangle u_{i+1,J} + (\langle C_{e},0\rangle + \langle-C_{w},0\rangle)u_{i,J} - \langle C_{w},0\rangle u_{i-1,J}$$

$$\tag{3.81a}$$

where

$$C_e = \rho \Delta y \frac{u_{i+1,J} + u_{i,J}}{2}, \tag{3.81b}$$

$$C_w = \rho \Delta y \frac{u_{i,J} + u_{i-1,J}}{2}. \tag{3.81c}$$

Here, the upwind difference method is adopted for discretization (recall (3.74a) and (3.74b) for the discretization by the upwind difference method). Similarly, the second term of the left hand side in (3.79) is

$$\int_{CV} \left(\rho u \frac{\partial u}{\partial x} \right) dV = -\langle -C_n, 0 \rangle u_{i,J+1} + (\langle C_n, 0 \rangle + \langle -C_s, 0 \rangle) u_{i,J} - \langle C_s, 0 \rangle u_{i,J-1} \tag{3.82a}$$

with

$$C_n = \rho \Delta x \frac{v_{I,j+1} + v_{I-1,j+1}}{2} \tag{3.82b}$$

$$C_s = \rho \Delta x \frac{v_{I,j} + v_{I-1,j}}{2} \tag{3.82c}$$

where the y-component of velocity, v, is treated as a known variable in order to gain a linear algebraic equation for the x-component of velocity, u. Discretization of the pressure term is relatively simple, since a pressure is directly defined at the center of boundary faces of the control volume for a staggered grid;

$$-\int_{CV} \left(\frac{\partial p}{\partial x} \right) dV = -\int\int \left(\frac{\partial p}{\partial x} \right) dx dy$$
$$= -\Delta y (p_e - p_w)$$
$$= -\Delta y (p_{I,J} - p_{I-1,J}). \tag{3.83}$$

The viscous terms are discretized as follows;

$$\int_{CV} \left(\mu \frac{\partial^2 u}{\partial x^2} \right) dV = \int\int \left(\mu \frac{\partial^2 u}{\partial x^2} \right) dx dy$$
$$= \Delta y \left[\left(\mu \frac{\partial u}{\partial x} \right)_e - \left(\mu \frac{\partial u}{\partial x} \right)_w \right]$$
$$= \Delta y \left[\mu \frac{u_{i+1,J} - u_{i,J}}{\Delta x} - \mu \frac{u_{i,J} - u_{i-1,J}}{\Delta x} \right]$$
$$= \mu \frac{\Delta y}{\Delta x} u_{i+1,J} - 2\mu \frac{\Delta y}{\Delta x} u_{i,J} + \mu \frac{\Delta y}{\Delta x} u_{i-1,J} \tag{3.84}$$

and

$$\int_{CV} \left(\mu \frac{\partial^2 u}{\partial y^2} \right) dV = \iint \left(\mu \frac{\partial^2 u}{\partial y^2} \right) dx dy$$

$$= \Delta x \left[\left(\mu \frac{\partial u}{\partial x} \right)_n - \left(\mu \frac{\partial u}{\partial x} \right)_s \right]$$

$$= \Delta x \left[\mu \frac{u_{i,J+1} - u_{i,J}}{\Delta y} - \mu \frac{u_{J,j} - u_{i,J-1}}{\Delta y} \right]$$

$$= \mu \frac{\Delta x}{\Delta y} u_{i,J+1} - 2\mu \frac{\Delta x}{\Delta y} u_{i,J} + \mu \frac{\Delta x}{\Delta y} u_{i,J-1}. \qquad (3.85)$$

Substituting all discretized results into the original equation (3.79), we obtain a discretized formula for the Navier–Stokes equation in the x-direction;

$$- \langle -C_e, 0 \rangle u_{i+1,J} + (\langle C_e, 0 \rangle + \langle -C_w, 0 \rangle) u_{i,J} - \langle C_w, 0 \rangle u_{i-1,J}$$
$$- \langle -C_n, 0 \rangle u_{i,J+1} + (\langle C_n, 0 \rangle + \langle -C_s, 0 \rangle) u_{i,J} - \langle C_s, 0 \rangle u_{i,J-1}$$
$$= -\Delta y (p_{I,J} - p_{I-1,J})$$
$$+ \mu \frac{\Delta y}{\Delta x} u_{i+1,J} - 2\mu \frac{\Delta y}{\Delta x} u_{i,J} + \mu \frac{\Delta y}{\Delta x} u_{i-1,J}$$
$$+ \mu \frac{\Delta x}{\Delta y} u_{i,J+1} - 2\mu \frac{\Delta x}{\Delta y} u_{i,J} + \mu \frac{\Delta x}{\Delta y} u_{i,J-1}. \qquad (3.86)$$

After some rearrangements, we gain

$$a_p u_{i,J} = a_e u_{i+1,J} + a_w u_{i-1,J} + a_n u_{i,J+1} + a_s u_{i,J-1} + S_p$$
$$a_e = \langle -C_e, 0 \rangle + \mu \frac{\Delta y}{\Delta x}, a_w = \langle C_w, 0 \rangle + \mu \frac{\Delta y}{\Delta x},$$
$$a_n = \langle -C_n, 0 \rangle + \mu \frac{\Delta x}{\Delta y}, a_s = \langle C_s, 0 \rangle + \mu \frac{\Delta x}{\Delta y}$$
$$a_p = a_e + a_w + a_n + a_s, S_p = \Delta y (p_{I-1,J} - p_{I,J}). \qquad (3.87)$$

The FVM formulation for the Navier–Stokes equation in the y-direction is accomplished in a similar fashion. The control volume and variables used for the FVM formulation of the Navier–Stokes equation for the y-direction is shown in Fig. 3.15.

Exercise 3.5 Formulate and discretize the Navier–Stokes equation in the y-direction by FVM.

3.5.2.3 SIMPLE Method: Coupling of the Equation of Continuity with the Navier–Stokes Equation

In the previous section, the Navier–Stokes equations were formulated as linear algebraic equations for velocities u and v. Now we need an equation for pressure, p. The rest of equation we have is the equation of continuity. However, it does not

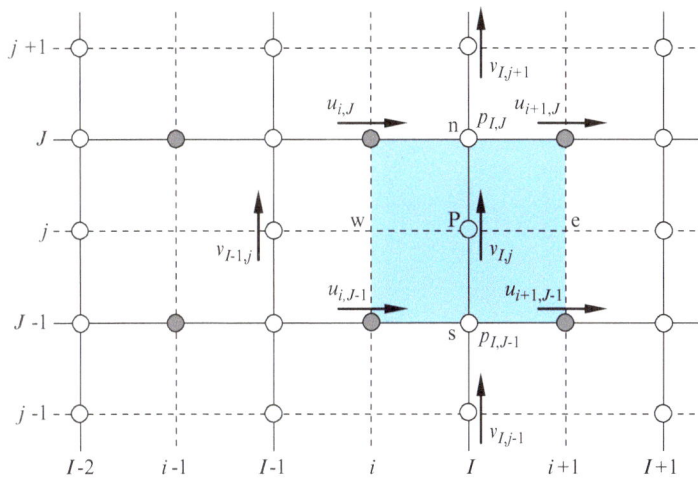

Fig. 3.15 Control volume for the Navier–Stokes equation in the y-direction

contain a pressure term. We therefore need to couple the equation of continuity with the Navier–Stokes equations in order to create the equation for pressure.

There are various ways to couple the equation of continuity with the Navier–Stokes equations. Of those, here we introduce so-called SIMPLE method originally developed by Patankar and Spalding (1972). For further explanations, we rewrite the algebraic equation for velocity u (3.87), as

$$a_e u_e = \sum a_{nb} u_{nb} + \Delta y(p_P - p_E). \tag{3.88a}$$

Similarly, we rewrite the algebraic equation for velocity v as

$$a_n v_n = \sum a_{nb} v_{nb} + \Delta x(p_P - p_N). \tag{3.88b}$$

Suffixes in (3.88a) and (3.88b) correspond to directions with respect to point P, the center of a control volume for the equation of continuity as illustrated in Fig. 3.16.

In (3.88a) and (3.88b), $\sum a_{nb} u_{nb}$ means a summation of a component $a_{nb} u_{nb}$ over all directions nb = E, W, S and N. Provided an estimate to pressure, p^*, we obtain approximates for velocities, u^* and v^* by

$$a_e u_e^* = \sum a_{nb} u_{nb}^* + \Delta y(p_P^* - p_E^*), \tag{3.89a}$$

$$a_n u_n^* = \sum a_{nb} v_{nb}^* + \Delta x(p_P^* - p_N^*). \tag{3.89b}$$

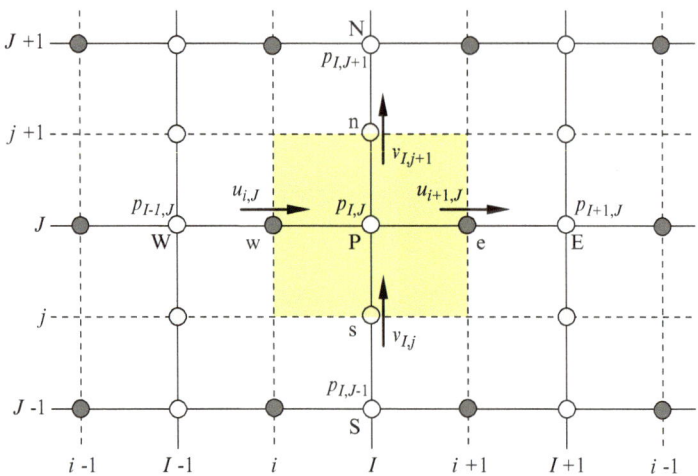

Fig. 3.16 Control volume for the equation of continuity

If correct values of pressure and velocities are p, u and v, correctors p', u' and v' against estimates p^*, u^* and v^* are expressed as

$$p = p^* + p',\tag{3.90a}$$

$$u = u^* + u',\tag{3.90b}$$

$$v = v^* + v'.\tag{3.90c}$$

Substituting these three equations into (3.88a) and (3.88b) and subtracting them from (3.89a) and (3.89b), we get

$$a_e u'_e = \sum a_{nb} u'_{nb} + \Delta y (p'_P - p'_E),\tag{3.91a}$$

$$a_n v'_n = \sum a_{nb} v'_{nb} + \Delta x (p'_P - p'_N).\tag{3.91b}$$

Assuming that the effect of correction from neighboring points is small in comparison to that from pressure, we drop the first term $\sum a_{nb} u'_{nb}$ or $\sum a_{nb} v'_{nb}$ in the right hand side. Note that omission of those does not affect the final solution because all corrections are zero when converged. Corrections for velocities are now written as

$$u'_e = \frac{\Delta y}{a_e} (p'_P - p'_E),\tag{3.92a}$$

$$v'_n = \frac{\Delta x}{a_n}(p'_P - p'_N). \tag{3.92b}$$

Substituting those into (3.91a) and (3.91b), we have

$$u_e = u_e^* + \frac{\Delta y}{a_e}(p'_P - p'_E), \tag{3.93a}$$

$$v_n = v_n^* + \frac{\Delta x}{a_n}(p'_P - p'_N). \tag{3.93b}$$

By the way, FVM formulation of the equation of continuity (3.75) over its control volume yields

$$\int_{CV}\left(\frac{\partial u}{\partial x} + \frac{\partial v}{\partial y}\right)dV = \int\int\left(\frac{\partial u}{\partial x} + \frac{\partial v}{\partial y}\right)dxdy$$
$$= \Delta y\int_w^e \frac{\partial u}{\partial x}dx + \Delta x\int_s^n \frac{\partial v}{\partial y}dy$$
$$= (u_e - u_w)\Delta y + (v_n - v_s)\Delta x = 0. \tag{3.94}$$

Substitution of (3.93a) and (3.93b) as well as analogous equations for u_w and v_s into this equation gives

$$a_P p'_P = a_E p'_E + a_W p'_W + a_N p'_N + a_S p'_S + b$$
$$a_E = \frac{(\Delta y)^2}{a_e}, \quad a_W = \frac{(\Delta y)^2}{a_w}, \quad a_N = \frac{(\Delta x)^2}{a_n}, \quad a_S = \frac{(\Delta x)^2}{a_s},$$
$$a_P = a_E + a_W + a_N + a_S,$$
$$b = \left(u_e^* - u_w^*\right)\Delta y + \left(v_n^* - v_s^*\right)\Delta x \tag{3.95}$$

By solving (3.95), we can calculate corrector p', and subsequently gain new p, u and v from (3.90a), (3.90b) and (3.90c). A series of this procedure is called SIMPLE method (Semi-Implicit Method for Pressure-Linked Equation). In this method, some terms are omitted when obtaining velocity corrections, which finally affects the formula of pressure correction equation. Other methods such as SIMPLER (Patankar 1980) and SIMPLEC (Van Doormal and Raithby 1984) use different treatments for pressure corrections.

3.6 Lagrange-Based Computational Fluid Dynamics

In Sect. 3.3, we introduced Euler-based methods of CFD. The formulation of Euler-based methods of CFD relies on a grid or mesh. However, generation of meshes is often difficult for complex geometries and phenomena showing large deformations in geometry. Recently, mesh-free methods of CFD have gained attentions of scientists who have been involved in the study of free-surface problems, multiphase problems and fluid–structure interaction problems. Of the mesh-free methods, a Lagrange-based method is called a particle method. The particle method represents fluid as a collection of particles and describes the motion of fluid as dynamic movements of particles. Each particle represents either a part of fluid or a part of physical objects that exist in the domain of interest. Thus, it possesses physical quantities such as density, viscosity, velocity, pressure and so forth. Governing equations are rewritten so as to trace each particle while conserving nature of continuum. In this section, we introduce one of the algorithms of the particle method, so-called moving particle semi-implicit method (MPS method) developed by Koshizuka and Oka (1996).

3.6.1 Governing Equations

The governing equations for the particle method are slightly different from the ones used for the Euler-based CFD. Again, we assume an incompressible fluid for simplicity. The governing equations for an incompressible fluid are given by

$$\frac{\partial \rho}{\partial t} = 0 \tag{3.96}$$

$$\rho \frac{\partial \mathbf{u}}{\partial t} = -\nabla p + \mu \nabla^2 \mathbf{u} + \mathbf{f} \tag{3.97}$$

where (3.96) is the equation of continuity that describes the conservation of mass, and (3.97) is the Navier–Stokes equation that describes the conservation of momentum. If you compare (3.96) with (3.31), you find that (3.96) does not have a divergence term of $\rho \mathbf{u}$. This is because the particle method is the Lagrange-based method in which we look at fluid motion with following an individual fluid parcel as it moves through space and time. As a consequence, the equation of continuity for the particle method is reduced to describing constancy of fluid density (or constancy of particle density). This is one of the features of the particle methods for an incompressible flow. In the particle method, if the number of particles remains constant and a mass of each particle does not change, a conservation of mass is automatically established. This in turn means that we need to impose a constraint on the density of particles to account for the conservation of mass. A reader also finds that a convective term vanishes in the Navier–Stokes equations. The reason for this is just the same as above. In the Lagrange-based CFD, the convective terms are eliminated from the material derivative.

3.6.2   Modeling of the Interaction Between Particles

Modeling of an interaction between particles is a key of the particle method. The way of modeling the interaction between particles is classified into a probabilistic model and a deterministic model. Main representatives of the probabilistic model include molecular dynamics, Monte Carlo method, lattice gas automaton (Frisch et al. 1987) and lattice Boltzmann equation. Because a fluid is continuum, use of the probabilistic model such as a random walk model for the interaction between particles requires a vast amount of particles to be simulated. In addition, a statistical process is needed to smoothen stochastic motions of particles for expressing continuum nature of fluid. Therefore, the probabilistic model is not suitable for describing flow. The MPS method described below adopted the deterministic model. In this model, once the initial and boundary conditions are provided, movements of particles are precisely determined by the equation of continuity, the Navier–Stokes equation and a particle interaction model.

The particle interaction model is required to represent partial derivative terms that appear in the governing equations. In other words, the particle interaction model is somehow equivalent to a spatial discretization for the Euler-based CFD.

In the MPS method, the particle interaction is expressed in terms of a weighting function, $w(r)$, defined by

$$w(r) = \begin{cases} r_e/r - 1 & \text{if } r < r_e \\ 0 & \text{else} \end{cases} \tag{3.98}$$

where r is a distance between two particles and r_e is the threshold. A particle number density $\langle n \rangle_i$ at particle i is calculated from

$$\langle n \rangle_i = \sum_{j \neq i} w(|\mathbf{r}_j - \mathbf{r}_i|) \tag{3.99}$$

which takes a summation of the weighting function over all neighboring particles. A gradient of physical quantity ϕ between particles i and j is described as

$$\langle \nabla \phi \rangle_i = \frac{d}{n^0} \sum_{j \neq i} \left[\frac{\phi_j - \phi'_i}{|\mathbf{r}_j - \mathbf{r}_i|^2} (\mathbf{r}_j - \mathbf{r}_i) w(|\mathbf{r}_j - \mathbf{r}_i|) \right] \tag{3.100}$$

where n^0 is the particle density over the domain and d is the number of space dimension. Practically, n^0 is estimated as an average particle number density at the initial arrangement of particles. ϕ_i' is calculated by

$$\phi_i' = \min\left(\phi_i, \phi_j\right) \text{ for } j \in \left(r_j < r_e\right) \tag{3.101}$$

which means that we use the minimum value of ϕ for ϕ_i' in the neighboring particles within distance r_e including particle i. By use of the minimum value of ϕ for ϕ_i', the force generated between particles is always repulsive. This tricky treatment is useful for numerical stabilization (Koshizuka et al. 1998). Laplacian of physical quantity ϕ in the MPS method is expressed by

$$\langle \nabla^2 \phi \rangle_i = \frac{2d}{n^0 \lambda} \sum_{j \neq i} \left[\left(\phi_j - \phi_i\right) w\left(|\mathbf{r}_j - \mathbf{r}_i|\right) \right] \tag{3.102}$$

where λ is a constant necessary to match (3.102) with an analytical solution of the divergence of (3.100). It is given by

$$\lambda = \frac{\displaystyle\sum_{j \neq i} \left[w\left(|\mathbf{r}_j - \mathbf{r}_i|\right) |\mathbf{r}_j - \mathbf{r}_i|^2 \right]}{\displaystyle\sum_{j \neq i} \left[w\left(|\mathbf{r}_j - \mathbf{r}_i|\right) \right]} \tag{3.103}$$

Equation (3.102) is actually the model of a transient diffusion of ϕ which particle i has to neighboring particles according to the weighting function.

3.6.3 Algorithm of the MPS Method

The algorithm of the MPS method is similar to SMAC (Simplified Marker and Cell) method (Amsden and Harlow 1970) that uses a semi-implicit scheme (pressure is solved implicitly, but the rest is solved explicitly). On the basis of SMAC method, the governing equations are modified as

$$\left[\frac{\partial \rho}{\partial t}\right]^{k+1} = 0 \tag{3.104}$$

$$\mathbf{u}^{k+1} = \mathbf{u}^k + \Delta t \left[\frac{1}{\rho}(-\nabla p)^{k+1} + \left(v\nabla^2 \mathbf{u} + \mathbf{f}\right)^k \right] \tag{3.105}$$

in which v is the kinematic viscosity, a ratio of the dynamic viscosity to the density, a forward difference method is applied for discretizing a temporal term, and superscript k stands for a time step.

Suppose that positions $\mathbf{r}_i^k$, velocities $\mathbf{u}_i^k$ and pressure p_i^k of all particles at time step k, which may be the initial condition at time step 0, are given. First, we explicitly calculate the viscous terms and external forces in (3.105) using velocities $\mathbf{u}_i^k$, and predict velocities $\mathbf{u}^*$ and positions $\mathbf{r}^*$ by

$$\mathbf{u}^* = \mathbf{u}^k + \Delta t \left(\mu \nabla^2 \mathbf{u} + \mathbf{f} \right)^k, \tag{3.106}$$

$$\mathbf{r}^* = \mathbf{r}^k + \Delta t \mathbf{u}^*. \tag{3.107}$$

Followed by calculating a particle density n^* from new positions of particles, we solve a Poisson equation for the pressure given as

$$\left(\nabla^2 p \right)^{k+1} = -\frac{\rho}{(\Delta t)^2} \frac{n_i^* - n^0}{n^0}. \tag{3.108}$$

Discretizing the left hand side of (3.108) with the Laplasian model of (3.102) as

$$\left(\nabla^2 p \right)^{k+1} = \frac{2d}{n^0 \lambda} \sum_{j \neq i} \left[\left(p_j^{k+1} - p_i^{k+1} \right) w \left(\left| \mathbf{r}_j^* - \mathbf{r}_i^* \right| \right) \right], \tag{3.109}$$

we obtain a linear algebraic equation for pressure, p^{k+1} to be solved. Once p^{k+1} is gained, correction for velocities, $\mathbf{u}'$, is calculated from

$$\mathbf{u}' = -\frac{\Delta t}{\rho} \nabla p^{k+1}. \tag{3.110}$$

Finally, true values of velocities and positions of particles at time step $k + 1$ are obtained by

$$\mathbf{u}^{k+1} = \mathbf{u}^* + \mathbf{u}', \tag{3.111a}$$

$$\mathbf{r}^{k+1} = \mathbf{r}^* + \Delta t \mathbf{u}'. \tag{3.111b}$$

This process is repeated until the end of time step to capture evolution of a flow field.

By the way, a reader may wonder where (3.108) and (3.110) come from. Deviation of these two equations is given as follows. Subtraction (3.106) from (3.105) yields

$$\mathbf{u}^{k+1} = \mathbf{u}^* + \frac{\Delta t}{\rho} \left(-\nabla p \right)^{k+1}. \tag{3.112}$$

Because of (3.111a), (3.112) is rewritten to (3.110). Deviation of (3.108) is slightly difficult. The equation of continuity for a compressible fluid in the Lagrange description is given by

$$\frac{\partial \rho}{\partial t} + \rho \nabla \cdot \mathbf{u} = 0. \tag{3.113}$$

Since density is approximated by the particle number density n in the MPS method, (3.113) is rewritten as

$$\frac{\partial n}{\partial t} + n^0 \nabla \cdot \mathbf{u} = 0 \tag{3.114}$$

where ρ for the divergence term is rewritten as n^0 because ρ is constant. If correction of the particle number density is determined by that of velocity, (3.114) is further rewritten as

$$\frac{\partial n'}{\partial t} + n^0 \nabla \cdot \mathbf{u}' = 0. \tag{3.115}$$

Discretizing the first term with a forward difference method in time, writing a deviation in the particle number density from the true value as

$$n' = n^0 - n^* \tag{3.116}$$

and substituting it into (3.115), we get

$$\frac{n^0 - n^*}{\Delta t} + n^0 \nabla \cdot \mathbf{u}' = 0. \tag{3.117}$$

Substitution of (3.110) into (3.117) gives (3.108).

3.7   Applications of Flow Simulations to Biomechanical Problems

The fluid mechanical discipline has been applied to studying flow in biology and medicine. One of the scientific landmarks in the history of biofluid studies was presented by the group of Prof. C.G. Caro, Imperial College London. The achievements of his group and colleagues are summarized in the book "The Mechanics of the Circulation" (Caro et al. 2011). Now, it is generally accepted that hemodynamics somehow plays a critical role in pathophysiological regulations of vascular structure and function (Kamiya and Togawa 1980; Kamiya et al. 1988). To relate fluid mechanical factors to pathogenesis of the vascular diseases including initiation, progression and localization of vascular diseases such as atherosclerotic lesions and aneurysms, numerous studies on blood flows have been made in the last decades and a variety of mechanical indices based on a wall shear stress and its derivatives both in time and space have been proposed (Ku et al. 1985; Gonzalez et al. 1992; Rossitti 1998; Meng et al. 2007; Shimogonya et al. 2009). Although any proposed theories for the initiation and development of vascular diseases are not conclusive, it would be true that application of fluid mechanics to vascular pathologies significantly advanced life science that was considered to be quite apart from mechanical engineering disciplines.

Biofluid disciplines are also essential in cardiovascular and respiratory therapies, which include the design and development of medical devices. Applications of blood flow analysis include the design of artificial hearts, valves, bypass grafts and stents as well as preoperative planning of targeted drug delivery and surgery. The study of airflow is applied to dose estimates of inhaled drugs in the lungs, the design of targeted therapeutic aerosols to more effectively treat respiratory diseases such as asthma, and development of mechanical ventilators. Furthermore, biofluid disciplines are used in innovation and validation of diagnostic techniques such as pulse wave velocity which is a measure of arterial stiffness and propagation wave velocity which is a measure of a left ventricular diastolic function. More recently, biofluid disciplines are applied to customize hydrodynamic cellular environments inside bioreactor systems to meet the unique requirements of different cell types and to facilitate the maturation of tissue-engineered constructs.

The great advances in computer technology during the past decade have enabled us to analyze various biological and physiological phenomena that were previously impossible to study numerically. This section introduces five case studies of biofluids by numerical simulations.

3.7.1 Analysis of Blood Flow in the Aorta

Hemodynamics have a major influence on blood coagulation and thromboembolism (Sakariassen and Barstad 1993), endothelial cell structure and function (Nerem 1993), and the uptake and accumulation of low-density lipoproteins on the arterial wall (Niwa et al. 2004). Many numerical studies have attempted to relate hemodynamic factors, such as the wall shear stress (WSS) and its derivatives in both time and space, to atherosclerotic lesions and aneurysms (Kleinstreuer et al. 2001). Similar attempts have been made for the aorta. Most of these studies, however, adopted a simplified geometry of the aorta, and the most frequent and apparent simplification was to ignore the branching and tapering of the aorta to eliminate the difficulty in defining computational grids (Mori and Yamaguchi 2002; Morris et al. 2005).

In this study, the hemodynamics in the human aorta is explored by combining magnetic resonance imaging (MRI) measurements and computational fluid dynamics (CFD) simulations to study the effects of branching of the aorta on blood flow. MRI was used to define the geometry of a human aorta, from which aorta models with or without three branches and taper were constructed. Cine phase-contrast

MRI was used to acquire 3-D time-resolved velocities at the inlet and outlets of the aorta (Yokosawa et al. 2005).

Two-dimensional cine phase-contrast MRI with a 1.5-T MR system (Signa Infinity EchoSpeed with the Excite option; General Electric, Fairfield, CT, USA) was used to obtain a magnitude image to provide anatomical information and phase images for attaining flow velocities in three orthogonal directions. The MR parameters were determined using standard settings for clinical examinations: repetition time (TR) 33 ms, echo time (TE) 5.4–6.0 ms, velocity encoding range (VENC) 150 cm/s, flip angle 30°, slice thickness 5 mm, matrix 192 × 192, and field of view (FOV) 32 × 24 cm. The subject was an adult male volunteer with no history of cardiac disease. All measurements were taken while the subject was holding his breath after a maximum expiration. The series of 30 equidistant images per heartbeat interval were acquired under electrocardiogram (ECG) synchronization.

Three computational models of the aorta were prepared: a model without branches or taper, a model with branches but no taper, and a model with branches and taper. It was assumed that the cross sections perpendicular to the centerline of the aorta were circular. The diameters of the cross sections were set to 2.6 cm for the ascending aorta, 1.0 cm for the brachiocephalic artery, and 0.7 cm for the left carotid and left subclavian arteries. The tapering applied to the aortic arch was estimated by fitting the exponential equation

$$A(x) = A_0 \exp(Cx) \tag{3.118}$$

to the data set for the cross-sectional area of the aortic arch obtained from the MRI. Here, x is the distance from the aortic inlet along the centerline, A is the cross-sectional area of the aortic arch, A_0 is the cross-sectional area at the aortic inlet, and C is a constant. Using the least squares method, we obtained $A_0 = 6.75$ cm^2 and $C = -0.037$. The diameter from the exit of the aortic arch to the descending aorta was set to a constant value of 2.28 cm.

Blood was treated as an incompressible Newtonian fluid with a density of 1.03×10^3 kg/m^3 and a viscosity of 4.0×10^{-3} Pa s. Computations were performed using the commercial CFD program SCRYU ver. 2.11 (Software Cradle, Osaka, Japan), which uses a finite volume method. At the inlet, all three components of the velocity data measured with MRI were imposed after mapping the measured MR data onto the computational grid inlet using a coordinate transformation. The flow velocity at a time instant that was not measured was interpolated using a cubic spline function. Parabolic flow was applied at the outlet of the aortic branches. The distribution ratio of flow to the branches and descending aorta was determined from the MR data and fixed throughout the cardiac cycle. It was set such that 15% of the total aortic inflow went to the brachiocephalic artery, 10% to the left carotid artery, and 8% to the left subclavian artery. A nonslip condition was applied at the wall and a traction-free condition was imposed at the end of the descending aorta.

The flow patterns in the full aorta model were presented in Fig. 3.17. In early systole, flow was directed axially toward the exit of the aorta. In the models with

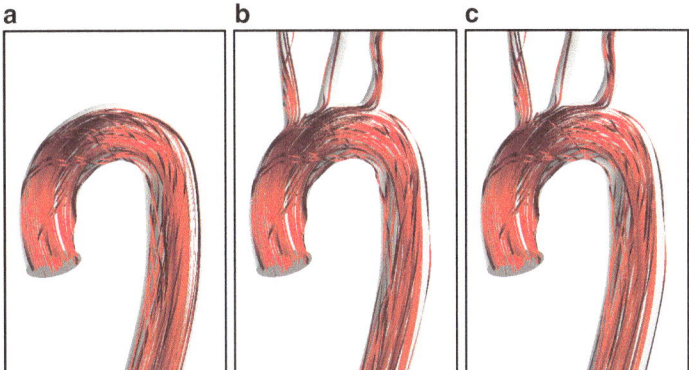

Fig. 3.17 Flow patterns at the peak systole. (**a**) The model without the branches and taper, (**b**) the model with branches but no taper and (**c**) the model with branches and taper

branches, some blood also flowed to the branches. From the peak of systole, a right-handed helical flow evolved from the ascending aorta, as shown in Fig. 3.17. Toward the end of systole, the axial flow diminished and the helical flow became more dominant. During diastole, no net flow was observed, although some regurgitant flow occurred at the branches.

The WSS and oscillatory shear index (OSI) (Buchanan and Kleinstreuer 1998) were evaluated as hemodynamics factors that might act on the arterial wall to cause vascular diseases. The WSS is calculated simply by multiplying a velocity gradient parallel to the wall with a viscosity. The OSI is defined by

$$\text{OSI} = \frac{1}{2}\left(1 - \left|\int_0^T \tau_w dt\right| \middle/ \int_0^T |\tau_w| dt\right). \tag{3.119}$$

Figure 3.18 shows the contour plots of the WSS at the peak of systole. The WSS distribution at the main trunk of the aorta was qualitatively similar. In all models, a relatively high WSS was found at the inner curvature of the proximal (the other side of this figure) and distal ends of the aortic arch, although the WSS for the model with the branches and no taper was slightly lower. For the models with branches, a high WSS occurred at the roots of the branches, particularly posteriorly. This tendency was found throughout systole. In diastole, the WSS was almost zero and was too small for comparison.

The contour plots of the OSI are illustrated in Fig. 3.19. The pattern was similar overall, although differences were observed at the roots of the branches and the inner curvature of the descending aorta. The presence of branches caused a high OSI at the roots of the branches and the left subclavian artery. The model with branches but no taper had a relatively high OSI region at the descending aorta.

The presence of branches caused the blood in the aortic arch to flow upward, giving rise to differences in the WSS and OSI at the roots of the branches. Since the

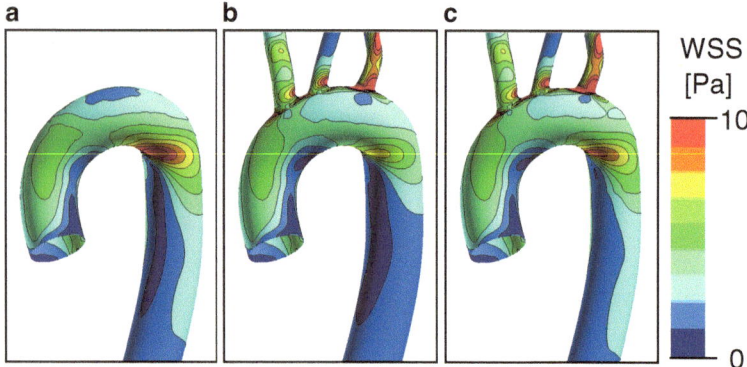

Fig. 3.18 WSS distribution at the peak systole. (**a**) The model without the branches and taper, (**b**) the model with branches but no taper and (**c**) the model with branches and taper

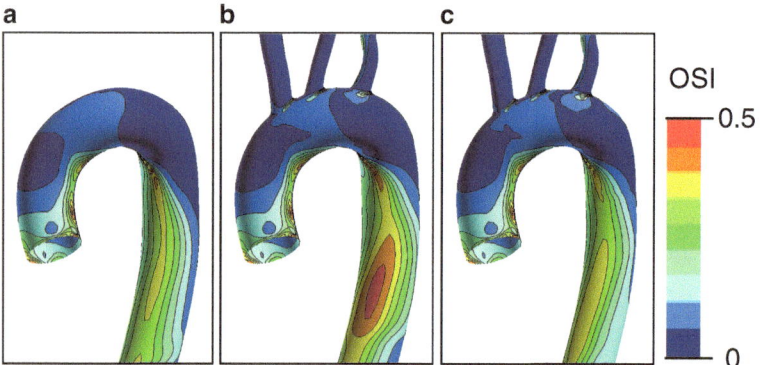

Fig. 3.19 OSI distribution at the peak systole. (**a**) The model without the branches and taper, (**b**) the model with branches but no taper and (**c**) the model with branches and taper

flow velocity is quite high in the aorta, the blood in the aorta cannot flow into the branches smoothly. As a result, flow separation occurred and the blood impinged on the posterior wall of the branching roots, elevating the WSS there. In addition, since the WSS was almost zero during diastole in all three models, the WSS fluctuated over a wider range and the OSI became higher at the roots of the branches.

Published numerical studies of the human aorta have included various simplifications of the geometry. The most apparent and frequent simplification was eliminating the aortic branches, and tapering was also often ignored. With these simplifications, the aorta was reduced to the simplest model that has neither branches and tapering. A comparison of the results obtained in this model with those obtained in the other models showed some differences in the hemodynamic factors at the roots of the branches and the inner curvature of the descending aorta. Although the affected regions were very limited, they were sites for aneurysms and atherosclerotic lesions (DeBakey et al. 1985). Therefore, from the perspective of biomedical engineering, it is important to include the branches and taper in models of the aorta to better predict the hemodynamics.

In conclusion of this section, the influence of geometric simplifications on the hemodynamics in the aorta was examined in a combined MRI and CFD study. Although the branches and taper did not affect the global patterns of flow in the aorta, the results showed their importance in analyses of the WSS and OSI, which are often of central interest for researchers who relate the aortic hemodynamics to vascular diseases, such as atherosclerotic lesions and aneurysms.

3.7.2   Analysis of Intraventricular Flow for the Assessment of a Left Ventricular Diastolic Function

The left ventricle (LV) distributes blood to a systemic circulation by contracting and expanding itself periodically. Whilst an LV function is often evaluated from a viewpoint of its contractility or a systolic function, it is reported that its diastolic function, defined as a function to fill blood without a compensatory increase of atrial filling pressures, also plays a key role in a blood pumping (Poleur et al. 1989). Recently, a method by using a color M-mode Doppler (CMD) echocardiography has been proposed for assessing LV diastolic function (De Mey et al. 2001). In this method, a velocity distribution along LV long axis spanning from the center of a mitral valve orifice to the ventricular apex is measured on the basis of a Doppler ultrasound technique. The measured velocity is presented in an M-mode image called a CMD echocardiogram where the magnitude of velocity is expressed by color and its brightness as depicted in Fig. 3.20. Although clinicians diagnose that a heart showing the pattern of Fig. 3.20b is worse than that of Fig. 3.20a, no mechanical evidence has been shown. Here we studied intraventricular flow during early diastole, aiming at an advance in diagnosis of the diastolic function of a left ventricle base on color M-mode Doppler echocardiograms.

The geometry of the left ventricular cavity at the end of diastole was defined using medical images so that it approximated the general anatomy of a human left ventricle. The ventricular volume was set to 120 cm^3 for a normal left ventricle and 180 cm^3 for an enlarged left ventricle, which represents dilated cardiomyopathy.

It is generally thought that deformation of the left ventricle is caused by spontaneous relaxation and contraction of myocardium. Clinical data on the magnitude of intraventricular blood pressure suggest that the blood pressure during diastole is not high enough to significantly deform the left ventricular wall (Sabbah and Stein 1981). Thus, we assumed that the movement of the ventricular wall was

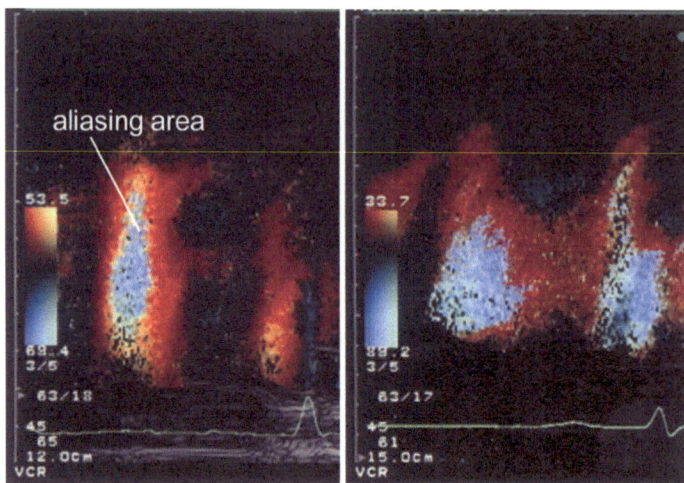

Fig. 3.20 Color M-mode Doppler echocardiogram. Normal (*left*) and diastolic dysfunction (*right*)

not affected by intraventricular flow dynamics. In addition, twisting and untwisting were neglected since their effects on intraventricular flow are relatively small (Nakamura et al. 2005). The velocity of the ventricular wall at the apex was determined to satisfy

$$V'(t) = \iint_S v_a(t)W(x, y, z)\, \mathbf{e} \cdot \mathbf{n}\mathrm{d}S \qquad (3.120)$$

where $V'(t)$ ($=\mathrm{d}V(t)/\mathrm{d}t$) is the time derivative of $V(t)$, $v_a(t)$ is the velocity of the ventricular apex, $W(x, y, z)$ is a weighting function of the moving velocity of the wall, $\mathbf{n}$ is a unit vector normal to the ventricular surface, and $\mathbf{e}$ is a unit vector parallel to the direction of wall movement. For the derivation of this equation, refer to Nakamura et al. (2003, 2004). If $V'(t)$ and $W(x, y, z)$ are provided, $v_a(t)$ is calculated from (3.120). Then, the velocity at any point on the wall is simply obtained from $v_a(t)$ and the weighting function $W(x, y, z)$. In this study, the weighting function $W(x, y, z)$ was set such that the wall at the base including the two valve orifices did not move, while the velocity of the wall increased going toward the apex.

The gradual opening of the mitral valve orifice is important for the formation of an intraventricular vortex (Nakamura et al. 2002). Therefore, the mitral valve was modeled as a planar, circular object with a core allowing blood to flow into the left ventricle. Hereafter, the core is referred to as the mitral valve orifice. The core was assumed to open or close axisymmetrically, and its size was changed as a function of the rate of volume change of the left ventricle (Nakamura et al. 2006).

Blood was treated as an incompressible Newtonian fluid with a density of $\rho = 1.05 \times 10^3$ kg/m^3 and a viscosity of $v = 3.5 \times 10^{-3}$ Pa s. Computations were performed under the moving boundary conditions using the CFD program ANSYS ver. 7.1 (Cybernet, Tokyo, Japan), which adopts a finite element method.

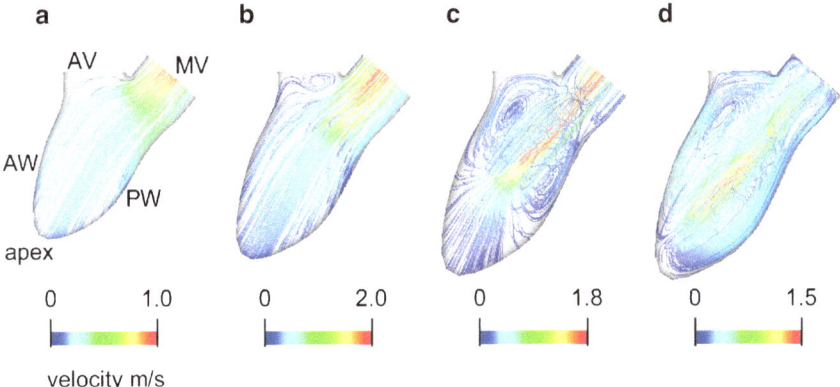

Fig. 3.21 Streamlines of blood flow for the left ventricle with a normal diastolic function. (**a**) $t = 0.06$ s, (**b**) $t = 0.12$ s, (**c**) $t = 0.18$ s, (**d**) $t = 0.24$ s. *MV* mitral valve, *AV* aortic valve, *AW* anterior wall, *PW* posterior wall

As boundary conditions, zero pressure and zero velocity were given to the opened and closed parts of the valve, respectively, and the velocity of the wall was applied to the ventricular wall based on a non-slip condition.

The volume change during diastole was determined from clinical data. The times from the beginning of diastole ($t = 0$) to the peak dV/dt and to the end of diastole were set to 0.12 and 0.24 s, respectively. The maximum of dV/dt in diastole was set so that a net change in the volume within the framework of early diastole, equivalent to the left ventricular diastolic function, was 20–70 cm^3 with an interval of 5 cm^3.

The simulation results showed the formation of an annular vortex under the aortic valve that was asymmetrically enlarged regardless of diastolic function. Figure 3.21 shows the flow patterns in the left ventricle with a net volume change of 60 cm^3. From the onset of diastole, blood flowed into the cavity through the mitral valve orifice to fill the ventricular cavity. The main flow headed toward the apex, while the other flows gradually diverged and headed toward the wall. Immediately after the peak of early diastole, the fluid elements under the aortic valve were induced to coil, forming a vortex. With further expansion, the vortex not only grew in size, but also extended in a circumferential direction, developing into an annular vortex that surrounded the blood inflow along the left ventricular long axis. At the mid-late stage of early diastole, another small vortex appeared in the space between the main flow heading straight toward the apex and the posterior wall. At this time, two asymmetric vortices were seen in the bisector plane. Toward the end of diastole, the annular vortex was amplified greatly. After the annular vortex was formed, the position of the fluid elements with a high velocity (the maximum velocity point) shifted toward the apex along the long axis, past the middle of the cavity.

The development of the vortex was retarded as the left ventricular diastolic function deteriorated. In this case, the annular vortex did not grow much larger, compared to the one observed in the normal left ventricle. The position of the maximum velocity point on the long axis did not shift toward the apex as much.

Fig. 3.22 Color M-mode
Doppler echocardiogram of a
ventricular filling flow along
the left ventricular long axis.
The magnitude of the velocity
is normalized by the
spatiotemporal maximum
velocity. (**a**) Volume
change $= 60 \text{ cm}^3$. (**b**)
Volume change $= 40 \text{ cm}^3$

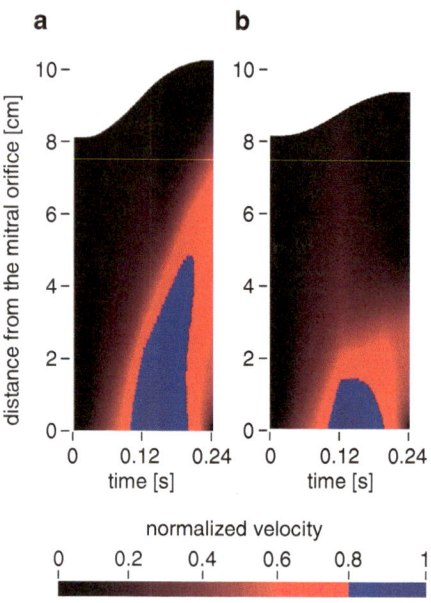

Fig. 3.23 Relationship
between the volume change
of the left ventricle and the
propagation velocity obtained
on the basis of Takatsuji's
method (1996)

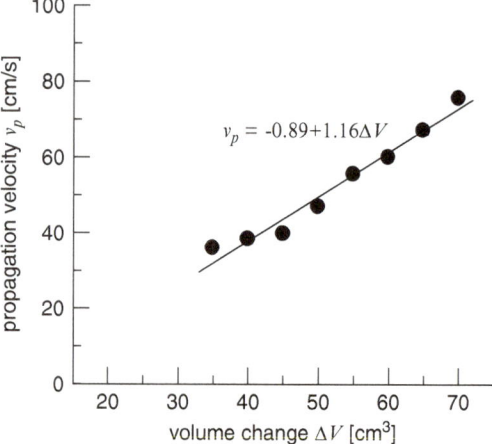

The difference in left ventricular diastolic function was clearly reflected in the
pattern of the CMD echocardiograms of the inflow velocity along the left ventricular
long axis, as illustrated in Fig. 3.22. The area that appears in blue in the middle of the
abscissa indicates the region where the velocity of fluid elements exceeded 80% of the
maximum velocity in the early diastolic phase and is called the aliasing area. More
quantitatively, the relationship between the left ventricular diastolic function and the
propagation velocity (Takatsuji et al. 1996) is plotted in Fig. 3.23. The flow propagation velocity increased linearly as the left ventricular diastolic function improved.

The fluidic mechanism required to bring about a change in the pattern of the CMD echocardiogram in accordance with the left ventricular diastolic function is controversial. Kawano et al. (2000) attributed it to a difference in the magnitude of the pressure gradient. In fact, a pressure gradient was formed from the base to the apex and the pressure propagated toward the apex. However, it occurred before the peak of early diastole, whereas the aliasing area elongated after the peak of early diastole. A detailed description of the intraventricular flow provided an opportunity to discuss the fluid mechanics factors that determine the shape of the aliasing area. At the beginning of diastole, the blood inflow through the mitral orifice diverged and its velocity decreased as it entered the main body of the left ventricle. This was due to the enlargement of the flow channel from the mitral orifice. As the annular vortex developed around the blood inflow along the left ventricular long axis, it narrowed the passage of the blood inflow. Consequently, the fluid elements surrounded by the annular vortex increased their velocities locally via the vena contracta effect, inducing the maximum velocity point at the same depth as the center of the annular vortex. Furthermore, since the center of the annular vortex moved toward the apex of the left ventricle as the vortex grew in size and increased in intensity, the maximum flow velocity propagated toward the apex. Therefore, the elongation of the aliasing area in the CMD echocardiogram is associated with the growth of the annular vortex toward the apex.

In general, the volume change during early diastole decreases as the left ventricular diastolic function deteriorates. The data in Takatsuji et al. (1996) suggest that the volume change of a left ventricle with low function during early diastole decreases by approximately 30% compared to one with normal diastolic function. If we assume that the net volume change for the normal left ventricle during early diastole is 60 cm^3, the left ventricle with the deteriorated diastolic function has a volume of 42 cm^3. According to this simulation, the respective traveling distance of the maximum velocity point and the flow propagation velocity for these cases are 4.8 cm and 60.7 cm/s for the normal left ventricle, and 2.5 cm and 39.8 cm/s for the deteriorated left ventricle. In other words, these indices decrease by 50% and 33%, respectively, in the deteriorated left ventricle. These results suggest that it is possible to detect a change in left ventricular diastolic function based on the CMD echocardiogram.

Computational fluid dynamics of the intraventricular flow during early diastole were used to investigate the relationship between left ventricular diastolic function and the pattern of the CMD echocardiogram. The findings suggest that a CMD echocardiogram, which expresses the spatiotemporal distribution of velocity along the long axis of the left ventricle, reflected the growth of an intraventricular annular vortex toward the ventricular apex during diastole, and the clinical evaluation of diastolic function of the left ventricle with this method indirectly captured the development of the intraventricular vortex.

3.7.3 Differentiation of Vascular Diseases by Pulse Wave Propagation

Pulse Wave Velocity (PWV) is recognized by clinicians as an index of the mechanical properties of human blood vessels (Ting et al. 1991; Thubrikar et al. 2001). PWV is generally determined by measuring a time delay between two waveforms which are recorded by plethysmography, and dividing the distance between the measurement points by the obtained time delay. However, the measured PWV of real human blood vessels will not always obey the Moens–Korteweg equation, which describes the PWV in ideal elastic tubes. Waveform analysis has been studied as an alternative diagnosis for cardiovascular disease, and reflected waves that occur in the diseased region may be a key for the estimation of the severity of disease. The study is aimed to explore how PWV changes in response to a local variation of vascular geometry that assumes vascular diseases (Fukui et al. 2007).

Two geometrical models having a local variation in geometry were created. Each model has a length of 1,000 mm, an internal radius of the normal part of 10 mm, and a wall thickness of the normal part of 2 mm. The diseased region, the length of which was 100 mm, was placed in the center of the model. One model has the central part of vessel narrowed by 70% in area to create stenosis (stenosed model). The other model has a local expansion of blood vessel represented with a Gaussian curve with the maximum radius of 25 mm (aneurysmal model).

The governing equations were the Navier–Stokes equations and the equation of continuity for the compressible fluid, and the equation of equilibrium for the solid. The solid was assumed to be a linear elastic material. The equations for the fluid were weakly coupled with those for solids; the fluid part and the structural part are solved sequentially and iteratively until the interface conditions are satisfied. Simulations were implemented with Radioss (MECALOG) that employs an arbitrary Lagrangean–Eulearian formulation. The boundary conditions used are follows. For fluids, a steady flow with Reynolds number 1,000 was imposed at the inlet of the artery as the basic flow, then a single rectangular pulse with Reynolds number 4,000 was imposed upon the basic flow to produce a propagating wave. At outlets, the 'silent boundary' condition, a modified equation of the Joukowsky equation (water hammer equation);

$$\frac{\partial p}{\partial t} = \rho^f c \frac{\partial u_n^f}{\partial t} + c \frac{(p_\infty - p)}{2l_c} \qquad (3.121)$$

was used to reduce reflections at the outlet. Here, superscript f (or s) stands for fluid (solid), n is the vector perpendicular to the cross section, p_∞ is the reference pressure, and l_c is a length characteristic of the grid size. At the interface between fluid and solid parts, a nonslip condition

$$\mathbf{u}^f = \mathbf{u}^s \qquad (3.122)$$

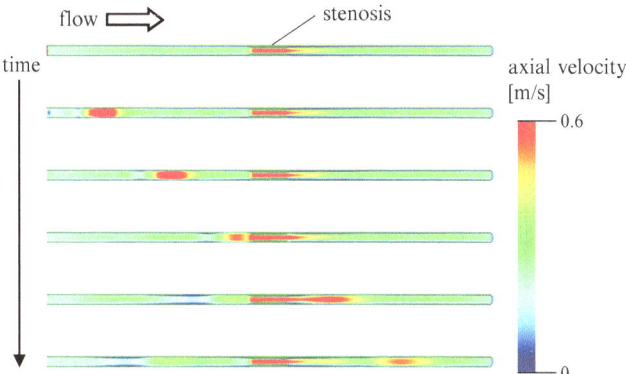

Fig. 3.24   A time series of the contour plot of the axial velocity at the interval of 0.05 s

was used. For solids, both ends of the channel were fixed as

$$\mathbf{u}^{s} = \omega^{s} = 0 \tag{3.123}$$

where ω is the angular velocity. The Young's modulus of the arterial wall was set to 0.5 MPa for both models. The viscosity coefficient of the blood was set to 4.0×10^{-3} Pa s.

Figure 3.24 shows the contours of the axial velocity v_y at the plane $x = 0$ of the stenosed artery. The high velocity portion propagated toward the periphery. The reflection was induced at the stenosis where characteristic impedance was changed. The characteristic impedances along the diseased arteries could be estimated by

$$Z = p/Q \tag{3.124}$$

where Z is the characteristic impedance, p is the pressure, and Q is the flow rate. In this study, the pressure p was obtained from the values in the center-line of the models, and the flow rate Q was calculated by the axial velocity in the center-line and the cross-sectional areas of the models. The characteristic impedance along the central axis was plotted in Fig. 3.25. In the stenosed artery, the characteristic impedance Z was suddenly decreased at the stenosis mainly due to the pressure drop. In the aneurysmal artery, the characteristic impedance was increased to the contrary.

Figure 3.26 shows the displacement of the wall Δr of the diseased arteries proximal to the diseased regions. The waves on the left hand travel peripherally, indicating these waves are fore-going. The waves on the right hand travel proximally, indicating these are back-going waves which occurred at the diseased regions.

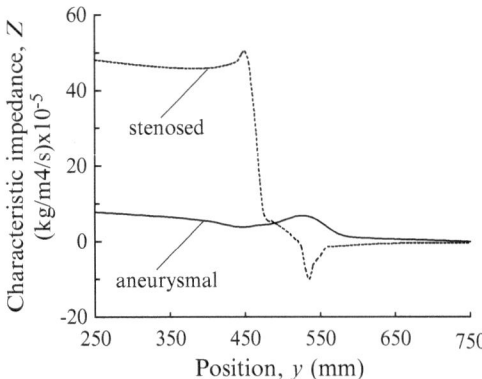

Fig. 3.25 Characteristic impedance of the stenosed and aneurysmal arteries

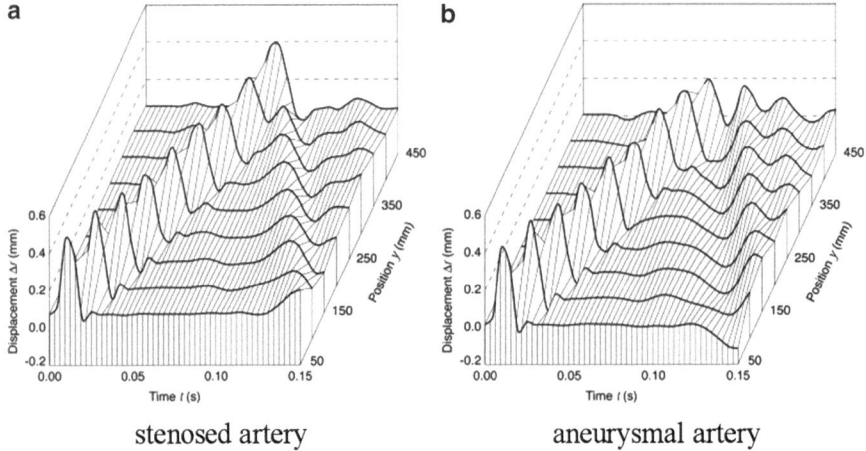

Fig. 3.26 A spatiotemporal variation of the wall displacement. (**a**) Stenosed artery and (**b**) aneurysmal artery

The phases of the back-going waves were opposite between the stenosed artery and the aneurysmal artery. This can be explained by

$$R_F = \frac{Z_0^{-1} - Z_d^{-1}}{Z_0^{-1} + Z_d^{-1}}$$

(3.125)

where R_F is the reflection coefficient, Z_0 is the characteristic impedance of the normal part of the artery, and Z_d is the characteristic impedance of the diseased region (Fung 1997). This indicates the phase of the reflected waves depends on whether the characteristic impedance of the diseased region is larger or smaller.

Therefore, the reflected waves from the stenosis and the aneurysm are different in their phase, and the wavelength of the reflected waves from the aneurysm is affected by the aneurysm length.

3.7.4   Primary Thrombus Formation by Platelet Aggregation

Platelets play an important role in hemostasis, responding to the change in blood flow when vessel wall is injured. While many physiological and biochemical reactions are involved in thrombogenesis, fluid mechanical factors such as fluid shear are important factors in the process (Ruggeri et al. 1999; Schmugge et al. 2003). This section describes a computer simulation method for the analysis of the formation of a primary thrombus due to platelet aggregation.

The particle method was used for the simulation of the primary thrombogenesis. The plasma and platelets were discretized into particles. Because the size of a platelet is considered to be small in comparison to the characteristic size of a blood vessel, the platelets were assumed to move along with blood plasma flow when they flow far from the injured wall. Once the platelets came into the vicinity of the injured wall within a distance of d_{ad}, they were induced to adhere. Motion of adhered platelets was assumed to be restricted by the adhesive force as long as they existed within the distance of d_{ad} from attachment point, although they were slightly drifted toward the downstream side. In this respect, the adhesive force implicitly includes platelet tether expansion. The adhesive force f_{ad} was expressed using a spring model in normal and tangential directions. The adhesive forces connected the platelet particle i to a particle j at the injured wall. Mathematically, the adhesive force $f_{ad,N}$ in a normal direction and $f_{ad,T}$ in a tangential direction are given by the following equations:

$$f_{ad,N} = k_{ad,N}(|r_{ij}| - r_0)n_{ij} \qquad \text{if } |r_{ij}| \leq d_{ad} \qquad (3.126a)$$

$$f_{ad,N} = \int k_{ad,N}(u_j - u_i) \cdot t_{ij}t_{ij}dt \quad \text{if } |r_{ij}| \leq d_{ad} \qquad (3.126b)$$

where $k_{ad,N}$ and $k_{ad,T}$ are the spring constants in normal and tangential directions, respectively; r_0 is the natural length of the normal spring; u_i and u_j are the velocities of particle i and j, respectively; and n_{ij} ($= |r_{ij}|/r_{ij}$) and t_{ij} are the normal and tangential vectors between the platelet particle i and j, respectively. The same model is used to represent an aggregating force of platelets f_{ag} whose magnitude is determined by the spring constants $k_{ag,N}$ and $k_{ag,T}$. This force acts between an adhered platelet particle i and a neighboring platelet particle j within a distance of d_{ag} from the particle i. Here, platelets and plasma never penetrate the injured wall. The adhesive forces $f_{ad} = f_{ad,N} + f_{ad,T}$ and aggregating forces $f_{ag} = f_{ag,N} + f_{ag,T}$

Fig. 3.27 Snapshots of the
formation process of a
primary thrombus at Re of
0.02

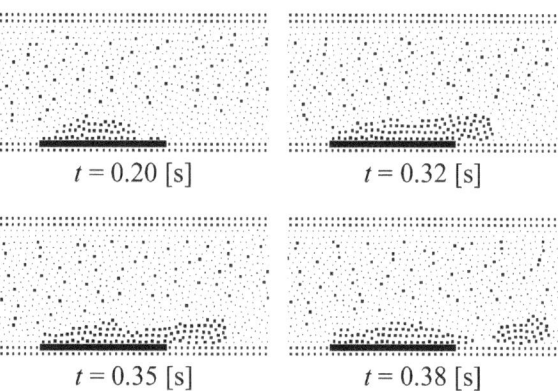

that act on this particle, are calculated and substituted into the external force f
of the Navier–Stokes equations. For further details, refer to Kamada et al.
(2010, 2011).

A two-dimensional blood flow simulation between parallel plates was conducted
for the case of the Reynolds number of 1.5×10^{-2}, 2.0×10^{-2} and 2.5×10^{-2}.
The sizes of the model were $L = 70$ μm in regard to axial flow length and $D = 20$
μm in regard to the distance between the plates. Particle distance was set to 1.0 μm
and the total number of particles was 3,044.

The formation process of primary thrombus at Re of 2.0×10^{-2} is shown in
Fig. 3.27. At the beginning of the simulation, platelets start to aggregate to the
injured wall due to the attractive force ($t = 0.2$ s), later developing into a cluster-
like thrombus ($t = 0.32$ s). As the height of the thrombus becomes larger with time,
the thrombus is deformed due to an increasing fluid force by plasma ($t = 0.35$ s).
Eventually, the thrombus collapses and moves downstream ($t = 0.38$ s). Figure 3.28
shows time course of the change in the number of platelets adhered to the injured
wall until the thrombus collapses. The number of the adhered platelets increases
monotonously with time. The increasing rate of the platelets is larger when Re is larger.

The simulation results suggest that the increase in the velocity of blood flow has
two different influences on thrombogenesis. One is (1) the increase in the number of
the platelets supplied by blood flow, which promotes the platelet aggregation. The
other is (2) the increase in fluid force acting on platelets against the attractive
and spring forces, which inhibits transportation of platelets to the injured wall.
Therefore, the change in the number of adhered platelets shown in Fig. 3.28 is the
result from the combined effects of these factors (1) and (2). For example,
the increasing rate of the number of the adhered platelets becomes larger when
the Re increases, in which the factor (1) is more effective than (2), whereas
the time for the thrombus collapse becomes longer, vice versa. These results
demonstrate the potential of the proposed particle method for the analysis of the
generation, growth and destruction of a primary thrombus.

Fig. 3.28 Time course of the number of adhered platelets to the injured wall

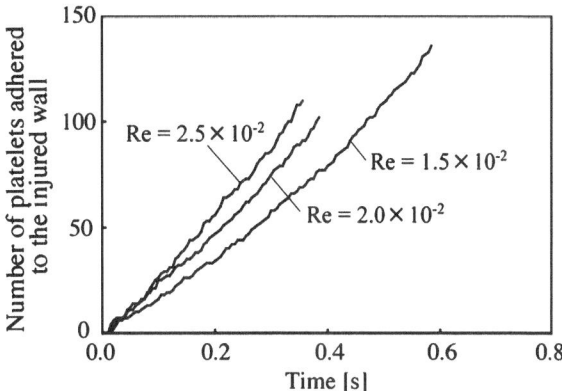

3.7.5   Analysis of the Behavior of Embolic Agents for Pre-operation Planning of Transcatheter Embolization

Transcatheter arterial embolization (TAE) is a way of occluding blood vessels such as arteriovenous fistula that are physiologically unnecessary. Although this technique becomes more popular in clinical practice for its minimal invasiveness, it is sometimes fraught with difficulties especially when a catheter cannot be placed directly in a target vessel. The risks associated with TAE may be reduced by better understanding of flowing behaviors of embolic agents in blood vessels. In this present study, a numerical model to simulate TAE is established to investigate influences of injection positions and intervals of spherical embolic agents (SEAs) on their flowing behaviors (Hidaka et al. 2008).

Flows of blood and SEAs were modeled by the equation of continuity along with the Navier–Stokes equation, re-formulated based on the MPS method. An SEA was modeled as an aggregate of particles. In brief, the particles consisting of the membrane of SEA were linked by springs which resist to stretching and bending. Here, stretching and bending energies W_S and W_b generated by deformation of SEA were described as

$$W_s = \frac{1}{2}k_s \sum_{i=1}^{N}\left(\frac{l_i - l_{i0}}{l_{i0}}\right)^2 \tag{3.127}$$

$$W_b = \frac{1}{2}k_b \sum_{k=1}^{N}\tan^2\frac{\theta_k}{2} \tag{3.128}$$

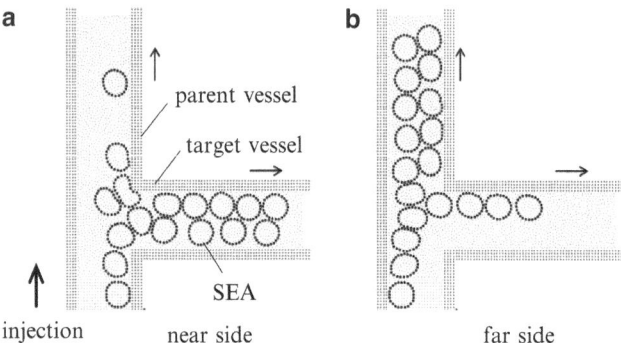

Fig. 3.29 Comparison of the flowing behaviors of SEAs when they are injected from (**a**) near side and (**b**) far side from the entrance of the target vessel

where l_{i0} and l_i are length of spring i at a natural state and after deformation, and θ_k is a bending angle of spring k. Moreover, energy for an area change of SEA was given as

$$W_a = \frac{1}{2}k_a\left(\frac{A - A_0}{A_0}\right)^2. \tag{3.129}$$

Based on the minimum energy principle, a force generated by deformation of SEA was calculated by

$$\mathbf{f} = -\frac{\partial(W_s + W_b + W_a)}{\partial \mathbf{r}} \tag{3.130}$$

which was substituted into a force term in the Navier–Stokes equations to accomplish a strong coupling simulation of flow and SEA deformation. See Chap. 4 for a further understanding of this modeling. The behavior of SEAs was examined in a T-junction blood vessel in which where a target vessel to be occluded with SEAs is connected perpendicularly to a main vessel.

The simulation showed that a distribution ratio of SEAs between the main and target vessels did not always coincide with a distribution ratio of blood flow evaluated in the absence of SEAs. Figure 3.29 provides snapshots of SEAs that were injected from different positions of the inlet. As seen, SEAs were prone to flow into the target blood vessel when injected from a proximal side of the target vessel, demonstrating that the radial position of injection significantly affected the flowing behavior of SEAs.

Simulation results with various injection frequencies are presented in Fig. 3.30. The injection frequency T was expressed with

$$T = \frac{u_0\delta}{d} \tag{3.131}$$

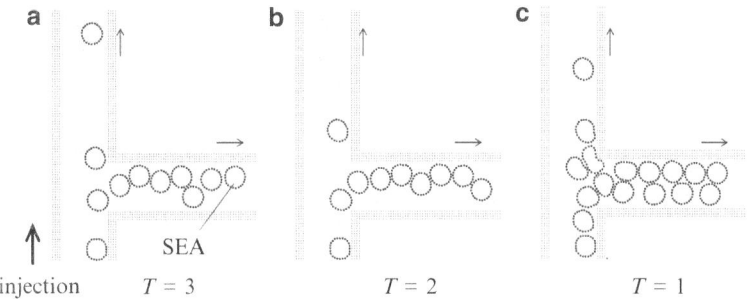

Fig. 3.30 Comparison of the flowing behaviors of SEAs when they are injected at various injection frequencies. (**a**) $T = 3$, (**b**) $T = 2$, and (**c**) $T = 1$

where d is the diameter of SEA, u_0 is the injection velocity and δ is an injection interval. Frequent injection of SEAs induced jam of SEAs in the target vessel, causing undesired flowing of SEAs into the main vessel. These results indicated the significant influence of injection position and interval on flowing behavior of SEAs.

In conclusion, the simulation results suggest injection positions and injection frequencies of SEAs have a significant impact on behaviors of SEAs, addressing the importance of controlling of SEAs injection in TAE. Future studies will incorporate real elastic properties of SEA now used in TAE on the basis of material tests (Hidaka et al. 2010).

3.8 Summary

This section provides a series of knowledge from fundamental theory of fluid mechanics to its applications to the study of biofluids. Although the study of biofluids has been progressed drastically for the last decades, there are many unsolved problems. Even just in blood flow analyses, we encounter various difficulties. They are, firstly, the complex geometry of the vascular system. It is no longer a simple tubular system with a circular cross section, but it is with curves, torsions, branches, tapers, etc. with individually different topological (i.e. connections of branches) structure. The vascular system differs from one person to other just like as their faces. Secondly, we always have to analyze it under unsteady conditions due to pulsating motion of the heart, the only energy source in the cardiovascular system. Thirdly, the wall of the artery is not rigid, and this imposes serious difficulty when it is combined with the unsteady flows. Fourthly, the blood, the working fluid of the cardiovascular system, is never a Newtonian fluid, since it consists of substantial volume fraction of cellular components such as the red blood cells. All these problems are nonlinear and difficult to deal with by experimental means in the real blood flow. This is particularly true when they are to be examined in a human body. Restriction of the measurements in the real human body is very strong and the CFD method is, in a sense, the only solution to analyze the blood flow phenomena.

References

Amsden AA, Harlow FH (1970) The SMAC Method: a numerical technique for calculating incompressible fluid flows. LA-4370, Los Alamos Scientific Laboratory, New Mexico

Baguchi P (2007) Mesoscale simulation of blood flow in small vessels. Biophys J 92:1858–1877

Baguchi P, Johnson PC, Popel AS (2005) Computational fluid dynamic simulation of aggregation of deformable cells in a shear flow. J Biomech Eng 127:1070–1080

Boryczko K, Dzwinel W, Yuen DA (2003) Dynamical clustering of red blood cells in capillary vessels. J Mol Model 9:16–33

Buchanan JR Jr, Kleinstreuer C (1998) Simulation of particle-hemodynamics in a partially occluded artery segment with implications to the initiation of microemboli and secondary stenosis. J Biomech Eng 120:446–454

Caro CG, Fitz-Gerald JM, Schroter RC (1969) Arterial wall shear and distribution of early atheroma in man. Nature 223:ll59–ll61

Caro CG, Pedley TJ, Schroter RC, Seed WA (2011) The mechanics of the circulation. Cambridge University Press, New York

De Mey S, De Sutter J, Vierendeels J, Verdonck P (2001) Diastolic filling and pressure imaging: taking advantage of the information in a colour M-mode Doppler image. Eur J Echocardiogr 12:219–233

DeBakey ME, Lawrie GM, Glaeser DH (1985) Patterns of atherosclerosis and their surgical significance. Ann Surg 201:115–131

Dzwinel W, Boryczko K, Yuen DA (2003) A discrete-particle model of blood dynamics in capillary vessels. J Colloid Interface Sci 258:163–173

Frisch U, d'Humières D, Hasslacher B, Lallemand P, Pomeau Y, Rivet JP (1987) Lattice gas hydrodynamics in two and three dimensions. Complex Syst 1:649–707

Fukui T, Parker KH, Imai Y, Tsubota K, Wada S, Yamaguchi T (2007) Differentiation of stenosed and aneurysmal arteries by pulse wave propagation analysis based on a fluid–solid interaction computational method. Technol Health Care 5:79–90

Fung YC (1997) Biomechanics circulation, 2nd edn. Springer, New York, pp 155–159

Gonzalez CF, Cho YI, Ortega HV, Moret J (1992) Intracranial aneurysms: flow analysis of their origin and progression. Am J Neuroradiol 13(1):181–188

Goto S (2008) Blood constitution: platelet aggregation, bleeding, and involvement of leukocytes. Rev Neurol Dis 5:S22–S27

Harlow FH, Welch JE (1965) Numerical calculation of time-dependent viscous incompressible flow of fluid with free surface. Phys Fluids 8:2182–2189

Hidaka K, Nakamura M, Osuga K, Komizu M, Wada S (2008) Effects of injection position and interval on the fraction of embolic agents at a bifurcation. Trans Jap Soc Med Biol Eng 6:647–654 (in Japanese)

Hidaka K, Nakamura M, Osuga K, Komizu M, Miyazaki H, Wada S (2010) Elastic characteristics of microspherical embolic agents used for vascular interventional radiology. J Mech Behav Biomed Mater 3:497–503

Kamada H, Tsubota K, Nakamura M, Ishikawa T, Yamaguchi T (2010) A three-dimensional particle simulation of the formation and collapse of a primary thrombus. Int J Numer Meth Biomed Eng 26:488–500

Kamada H, Tsubota K, Nakamura M, Wada S, Ishikawa T, Yamaguchi T (2011) Computational study on effect of stenosis on primary thrombus formation. Biorheology 48:99–114

Kawano Y, Ohmori K, Wada Y, Kondo I, Mizushige K, Senda S, Nozaki S, Kohno M (2000) Anovel color M-mode Doppler echocardiographic index for left ventricular relaxation: depth of the maximal velocity point of left ventricular inflow in early diastole. Heart Vessels 15:205–213

Kamiya A, Togawa T (1980) Adaptive regulation of wall shear stress to flow change in the canine carotid artery. Am J Physiol 239:H14–21

Kamiya A, Ando J, Shibata M, Masuda H (1988) Roles of fluid shear stress in physiological regulation of vascular structure and function. Biorheology 25:271–278

Kleinstreuer C, Hyun S, Buchanan JR Jr, Longest PW, Archie JP Jr, Truskey GA (2001) Hemodynamic parameters and early intimal thickening in branching blood vessels. Crit Rev Biomed Eng 29:1–64

Koshizuka S, Oka Y (1996) Moving particle semi-implicit method for fragmentation of incompressible fluid. Nucl Sci Eng 123:421–434

Koshizuka S, Nobe A, Oka Y (1998) Numerical analysis of breaking waves using the moving particle semi-implicit method. Int J Numer Meth Fluid 26:751–769

Ku DN, Giddens DP, Zarins CK, Glagov S (1985) Pulsatile flow and atherosclerosis in the human carotid bifurcation. Positive correlation between plaque location and low oscillating shear stress. Arteriosclerosis 5:293–302

Liu Y, Liu WK (2006) Rheology of red blood cell aggregation by computer simulation. J Comput Phys 220:139–154

Meng H, Wang Z, Hoi Y, Gao L, Metaxa E, Swartz DD, Kolega J (2007) Complex hemodynamics at the apex of an arterial bifurcation induces vascular remodeling resembling cerebral aneurysm initiation. Stroke 38:1924–1931

Mori D, Yamaguchi T (2002) Computational fluid dynamics modeling and analysis of the effect of 3D distortion of the human aortic arch. Comp Meth Biomech Biomed Eng 5:249–260

Morris L, Delassus P, Callanan A, Walsh M, Wallis F, Grace P, McGloughlin T (2005) 3-D numerical simulation of blood flow through models of the human aorta. J Biomech Eng 127:767–775

Nakamura M, Wada S (2011) Mesoscopic blood flow simulation considering hematocrit-dependent viscosity. J Biomech Sci Eng 5:578–590

Nakamura M, Wada S, Mikami T, Kitabatake A, Karino T (2002) A computational fluid mechanical study on the effects of opening and closing of the mitral orifice on a transmitral flow velocity profile and an early diastolic intraventricular flow. JSME Int J Ser C 45:913–922

Nakamura M, Wada S, Mikami T, Kitabatake A, Karino T (2003) Computational study on the evolution of an intraventricular vortical flow during early diastole for the interpretation of color M-mode Doppler echocardiograms. Biomech Model Mechanobiol 2:59–72

Nakamura M, Wada S, Mikami T, Kitabatake A, Karino T, Yamaguchi T (2004) Effect of flow disturbances remaining at the beginning of diastole on intraventricular diastolic flow and colour M-mode Doppler echocardiograms. Med Biol Eng Comput 42:509–515

Nakamura M, Wada S, Karino T, Yamaguchi T (2005) Effects of a ventricular untwisting on intraventricular diastolic flow and color M-mode Doppler echocardiograms. Technol Health Care 13:269–280

Nakamura M, Wada S, Yamaguchi T (2006) Influence of the opening mode of the mitral valve orifice on intraventricular hemodynamics. Ann Biomed Eng 34:927–935

Nerem RM (1993) Hemodynamics and the vascular endothelium. J Biomech Eng 115:510–514

Niwa K, Kado T, Sakai J, Karino T (2004) The effects of a shear flow on the uptake of LDL and acetylated LDL by an EC monoculture and an EC-SMC coculture. Ann Biomed Eng 32:537–543

Patankar SV (1980) Numerical heat transfer and fluid flow. Hemisphere Publishing Corporation, Taylor & Francis Group, New York

Patankar SV, Spalding DB (1972) A calculation procedure for heat, mass and momentum transfer in three-dimensional parabolic flows. Int J Heat Mass Transfer 15:1787–1806

Poleur H, Hanet C, Gurne O, Rousseau MF (1989) Focus on diastolic dysfunction: a new approach to heart failure therapy. Br J Clin Pharmacol 28(Suppl 1):41S–52S

Rossitti S (1998) Shear stress in cerebral arteries carrying saccular aneurysms. A preliminary study. Acta Radiol 39:711–717

Ruggeri ZM, Dent JA, Saldiver E (1999) Contribution of distinct adhesive interactions to platelet aggregation in flowing blood. Blood 94:172–178

Sabbah HN, Stein PD (1981) Pressure-diameter relations during early diastole in dogs. Incompatibility with the concept of passive left ventricular filling. Circ Res 48:357–365

Sakariassen KS, Barstad RM (1993) Mechanisms of thromboembolism at arterial plaques. Blood Coagul Fibrinolysis 4:615–625

Schmugge M, Rand ML, Freedman J (2003) Platelets and von Willebrand factor. Transfus Apher Sci 28:269–277

Secomb TW (1991) Red blood cell mechanics and capillary blood rheology. Cell Biophys 18:231–251

Shimogonya Y, Ishikawa T, Imai Y, Matsuki N, Yamaguchi T (2009) Can temporal fluctuation in spatial wall shear stress gradient initiate a cerebral aneurysm? A proposed novel hemodynamic index, the gradient oscillatory number (GON). J Biomech 42:550–554

Sun C, Munn LL (2005) Particulate nature of blood determines macroscopic rheology: a 2-D lattice Boltzmann analysis. Biophys J 88:1635–1645

Takatsuji H, Mikami T, Urasawa K, Teranishi J, Onozuka H, Takagi C, Makita Y, Matsuo H, Kusuoka H, Kitabatake A (1996) A new approach for evaluation of left ventricular diastolic function: spatial and temporal analysis of left ventricular filling flow propagation by color M-mode Doppler echocardiography. J Am Coll Cardiol 27:365–371

Thubrikar MJ, Labrosse M, Robicsek F, Al-Soudi J, Fowler B (2001) Mechanical properties of abdominal aortic aneurysm wall. J Med Eng Technol 25:133–142

Ting CT, Chou CY, Chang MS, Wang SP, Chiang BN, Yin FC (1991) Arterial hemodynamics in human hypertension. Effects of adrenergic blockade. Circulation 84:1049–1057

Van Doormal JP, Raithby GD (1984) Enhancements of the SIMPLE method for predicting incompressible fluid flows. Numer Heat Transfer 7:147–163

Womersley JR (1955) Method for the calculation of velocity, rate flow, and viscous drag in arteries when the pressure gradient is known. J Physiol 127:553–563

Yokosawa S, Nakamura M, Wada S, Isoda H, Takeda H, Yamaguchi T (2005) Quantitative measurements on the human ascending aortic flow using 2D cine phase-contrast magnetic resonance imaging. JSME Int J Ser C 48:459–467

Zhang J, Johnson PC, Popel AS (2007) An immersed boundary lattice Boltzmann approach to simulate deformable liquid capsules and its application to microscopic blood flows. Phys Biol 4:285–295

Zhang J, Johnson PC, Popel AS (2008) Red blood cell aggregation and dissociation in shear flows simulated by lattice Boltzmann method. J Biomech 41:47–55

Zhang J, Johnson PC, Popel AS (2009) Effects of erythrocyte deformability and aggregation on the cell free layer and apparent viscosity of microscopic blood flows. Microvasc Res 77:265–272

Chapter 4
Spring Network Modeling Based on the Minimum Energy Concept

In the preceding chapters, biological elements were modeled as a continuum body and their mechanical behavior was described by solid and/or fluid mechanics. Biological systems consisting of diversely scaled elements from biomolecule to cells, tissues and organs, however, include various mechanical components which cannot be always modeled as a continuum body. For example, macroscopic stress and strain are not well defined in a cell where cytoskeletons are discretely distributed. Such a mechanical component often plays an important role in biological events including the response of a cell to a mechanical stimulation, remodeling and growth of tissue, aggregation of red blood cells in blood flow, and so on. In order to understand the underlying mechanics behind biology having the multi-scale system and how mechanics contributes to regulating biological functions, it is essential to develop modeling and simulation methodology to combine a continuum model with a discrete model on the basis of a unified principle of mechanics that works across all scales. This chapter describes fundamentals of a spring network model which can be integrated with the continuum model based on the minimum energy concept. The chapter also presents computational biomechanics analyses of the shape and mechanical behavior of a red blood cell and the mechanical properties of a eukaryotic cell using a spring network model. As further applications, a hybrid modeling of the spring network and continuum elements is introduced for the rule-based simulation of aneurysm development and the multi-scale simulation of blood flow.

Keywords Aneurysm • Blood flow • Cell mechanics • Mechano-cell model • Minimum energy principle • Motion equation • Multi-scale simulation • Red blood cell • Rule-based simulation • Spring network model • Tissue mechanics

M. Tanaka et al., *Computational Biomechanics*, A First Course in "In Silico Medicine" 3,   141
DOI 10.1007/978-4-431-54073-1_4, © Springer 2012

4.1 Fundamentals of Spring Network Mechanics

4.1.1 Single Spring Model

Consider a stretch of a linear spring with a spring constant, k_s, and a natural length, L_0, to which an external force, F, is applied as shown in Fig. 4.1. The relationship between the force and the deformation of the spring is expressed as

$$F = k_s x \ \text{ and } \ x = L - L_0 \tag{4.1}$$

where L is a length of the stretched spring. According to the conservation law of energy, work done by the external force is equal to the elastic energy stored in the spring if energy dissipation by friction is ignored. Then, the elastic energy is written as

$$W_s = \int_0^x F dx = \frac{1}{2} k_s x^2. \tag{4.2}$$

Using this elastic energy function of x, the spring force can be alternatively described as

$$f = -\frac{dW_s}{dx} \ \text{ and } F = -f \tag{4.3}$$

where f is an internal force generated in the spring, which is balanced with the external force, F. The plus or minus sign of the force indicates the direction of the force along the stretching direction. These relations hold for a compression of the spring in which the deformation of x is negative.

When a spring is located in a space, the relationship between force and deformation is described in a vector form. Define position vectors, $\mathbf{r}_1$ and $\mathbf{r}_2$, of nodal points at both edges of a spring as shown in Fig. 4.2. The internal forces at nodal points 1 and 2 are written as

$$\mathbf{f}_1 = k_s(L - L_0)\frac{\mathbf{r}_2 - \mathbf{r}_1}{L} \ \text{ and } \mathbf{f}_2 = k_s(L - L_0)\frac{\mathbf{r}_1 - \mathbf{r}_2}{L} \tag{4.4}$$

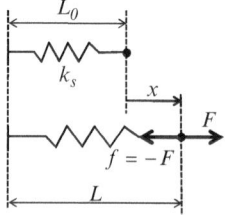

Fig. 4.1 Stretch of a linear spring

Fig. 4.2 General description
of a stretch of a linear spring

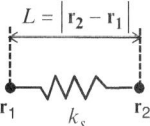

Fig. 4.3 Stretching of a
linear spring in a 2D space

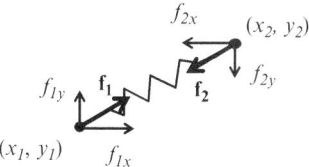

where

$$L = |\mathbf{r}_1 - \mathbf{r}_2|. \tag{4.5}$$

According to (4.3), the internal forces are simply expressed as

$$\mathbf{f}_i = -\frac{\partial W_s}{\partial \mathbf{r}_i} \tag{4.6}$$

using the elastic energy function of the position vectors, $\mathbf{r}_1$ and $\mathbf{r}_2$,

$$W_s = \frac{1}{2} k_s (L(\mathbf{r}_1, \mathbf{r}_2) - L_0)^2. \tag{4.7}$$

For example, as shown in Fig. 4.3, consider the x–y coordinate system in a 2-dimensional space given by $\mathbf{r}_1(x_1, y_1)$ and $\mathbf{r}_2(x_2, y_2)$. The length of a spring is

$$L = \sqrt{(x_2 - x_1)^2 + (y_2 - y_1)^2} \tag{4.8}$$

and the components of force vectors, $\mathbf{f}_1(f_{1x}, f_{1y})$ and $\mathbf{f}_2 \ (f_{2x}, f_{2y})$, are, respectively, obtained by

$$f_{1x} = -\frac{\partial W}{\partial x_1} = -\frac{dW}{dL}\frac{\partial L}{\partial x_1}, \ f_{1y} = -\frac{\partial W}{\partial y_1} = -\frac{dW}{dL}\frac{\partial L}{\partial y_1} \tag{4.9a}$$

and

$$f_{2x} = -\frac{\partial W}{\partial x_2} = -\frac{dW}{dL}\frac{\partial L}{\partial x_2}, \ f_{2y} = -\frac{\partial W}{\partial y_2} = -\frac{dW}{dL}\frac{\partial L}{\partial y_2} \tag{4.9b}$$

It is noted that the direction of force is automatically determined from (4.6).

Fig. 4.4 Network of linear
springs

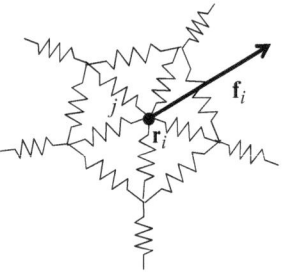

Exercise 4.1 Confirm that (4.9a) and (4.9b) are identical to (4.4) in the x–y coordinate system.

4.1.2   Network Spring Model

When springs are connected each other to form a network structure as shown in Fig. 4.4, the expression of (4.6) using an elastic energy function is useful to describe an internal force vector at nodal point, i. The total elastic energy of the network springs is obtained by summing an elastic energy stored in each spring, j, with a spring constant, k_j, and nodal points, j_1 and j_2, at both edges of the spring. Consequently the total elastic energy is expressed as a function of positional vectors of nodal points,

$$W_s(\mathbf{r}_1, \mathbf{r}_2, \ldots \mathbf{r}_N) = \sum_j W_{sj}(\mathbf{r}_{j_1}, \mathbf{r}_{j_2})$$
$$= \frac{1}{2} \sum_j k_{sj}(L_j(\mathbf{r}_{j_1}, \mathbf{r}_{j_2}) - L_{0j})^2. \qquad (4.10)$$

The internal force at the nodal point, i (1, 2, ..., N), is described by (4.6).

4.1.3   Bending Spring Model

In the previous sections, we considered a deformation of a linear spring and the force along the spring, which describes an in-plane deformation of a triangular element. An out-of-plane deformation of the element can be expressed by a bending spring connecting neighboring elements. Here we assume two triangular elements consisting of four nodal points as shown in Fig. 4.5 and introduce the bending energy as a function of a contacting angle between two elements, θ. It is written as

Fig. 4.5 Bending spring

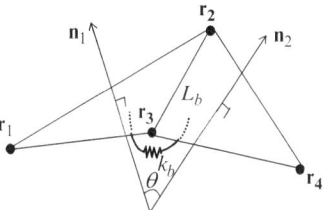

$$W_b = \frac{1}{2} k_b L_b \theta^2 \tag{4.11}$$

where k_b is a spring constant and L_b is a length of contacting edge. According to vector analysis,

$$L_b = |\mathbf{r}_2 - \mathbf{r}_3| \tag{4.12}$$

and

$$\theta = \arccos(\mathbf{n}_1 \cdot \mathbf{n}_2) \tag{4.13}$$

$$\mathbf{n}_1 = \frac{(\mathbf{r}_3 - \mathbf{r}_1) \times (\mathbf{r}_2 - \mathbf{r}_1)}{|(\mathbf{r}_3 - \mathbf{r}_1) \times (\mathbf{r}_2 - \mathbf{r}_1)|} \tag{4.14a}$$

$$\mathbf{n}_2 = \frac{(\mathbf{r}_2 - \mathbf{r}_4) \times (\mathbf{r}_3 - \mathbf{r}_4)}{|(\mathbf{r}_2 - \mathbf{r}_4) \times (\mathbf{r}_3 - \mathbf{r}_4)|} \tag{4.14b}$$

where $\mathbf{n}_1$ and $\mathbf{n}_2$ are normal unit vectors of two elements. Thus, the bending energy, W_b, is a function of position vectors at four nodal points and then the internal forces generated by the bending spring at nodal point, i ($=1, 2, 3,$ and 4), is obtained by

$$\mathbf{f}_i = -\frac{\partial W_b}{\partial \mathbf{r}_i}. \tag{4.15}$$

Exercise 4.2 When the contacting angle of θ is small, the energy function of (4.11) can be approximated as

$$W = \frac{1}{2} k_b L_b \sin^2 \theta \tag{4.16}$$

(1) Describe the energy as a function of $\mathbf{r}_1$, $\mathbf{r}_2$, $\mathbf{r}_3$, and $\mathbf{r}_4$.
(2) Obtain the x, y, and z components of the force vectors of $\mathbf{f}_1$, $\mathbf{f}_2$, $\mathbf{f}_3$, and $\mathbf{f}_4$.
(3) Confirm that $\mathbf{f}_1$ and $\mathbf{f}_4$ are normal to each element.
(4) Show that $\mathbf{f}_1 + \mathbf{f}_2 + \mathbf{f}_3 + \mathbf{f}_4 = \mathbf{0}$.

Fig. 4.6 Area of a triangular element

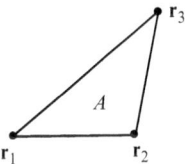

Fig. 4.7 Volume of tetrahedral element

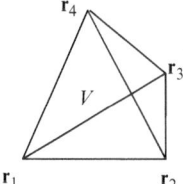

4.1.4   Extended Spring Model

By extending a spring model described in the energy form, the force can be derived from any energy functions written with position vectors of nodal points even if a spring does not explicitly appear in the model. For example, the area of a triangular element shown in Fig. 4.6 is written by the cross product as

$$A = \frac{1}{2}|(\mathbf{r}_2 - \mathbf{r}_1) \times (\mathbf{r}_3 - \mathbf{r}_1)| \tag{4.17}$$

using position vectors of three nodal points, $\mathbf{r}_1$, $\mathbf{r}_2$, and $\mathbf{r}_3$. Thus, if we define the energy function caused by a change in the area of a triangular element as

$$W_a = \frac{1}{2}k_a \left(\frac{A - A_0}{A_0}\right)^2 A_0 \tag{4.18}$$

where k_a is a constant and A_0 is a reference area of the element, the force vector at nodal point, i (=1,2, and 3), is obtained by

$$\mathbf{f}_i = -\frac{\partial W_a}{\partial \mathbf{r}_i}. \tag{4.19}$$

In a similar way, forces generated by a volume change of a tetrahedral element shown in Fig. 4.7 are obtained from the energy function,

$$W_V = \frac{1}{2}k_v \left(\frac{V - V_0}{V_0}\right)^2 V_0 \tag{4.20}$$

where the suffix 0 in (4.20) denotes a reference state, and V is a volume of the element and is written by the vector triple product as

$$V = \frac{1}{6}(\mathbf{r}_2 - \mathbf{r}_1) \times (\mathbf{r}_3 - \mathbf{r}_1) \bullet (\mathbf{r}_4 - \mathbf{r}_1). \tag{4.21}$$

If we take a natural state of an element in (4.18) and (4.20) as reference, the constants k_a and k_v are called the area expansion modulus and the bulk modulus in a continuum model, respectively. Equations (4.18) and (4.20) are also mathematically used as a penalty function to give constraint on the area and volume of the element. In this case, A_0 and V_0 are set to be a desired value, and k_a and k_v are weight coefficients.

Exercise 4.3

(1) Obtain components of force vectors generated by the energy function of (4.18) and (4.20) in the Cartesian coordinate.
(2) How does the element deform by these forces?
(3) Show that forces derived from (4.20) are equivalent to those generated by an inside pressure of the element, which is represented as $p = k_v (V - V_0)$.

4.1.5 Extension to Continuum Model

According to the solid mechanics, the strain energy density function of an elastic body is described as

$$dW_c = \frac{1}{2}\boldsymbol{\sigma}^T \boldsymbol{\varepsilon} \tag{4.22}$$

where $\boldsymbol{\sigma}$ and $\boldsymbol{\varepsilon}$ are stress and a strain tensors in vector form, respectively. A constitutive law of a material gives a stress–strain relationship (see Sect. 2.1.3). For a 2D linear elastic material with a constant thickness, h, the strain energy of a triangular element is given by

$$W_c = \frac{1}{2}hA\boldsymbol{\varepsilon}^T \mathbf{D}\boldsymbol{\varepsilon} \tag{4.23}$$

where A is the area of element, $\boldsymbol{\varepsilon} = (\varepsilon_x, \varepsilon_y, \varepsilon_{xy})^T$ is a strain tensor in vector form, and $\mathbf{D}$ is the elastic modulus tensor for plane strain. Furthermore, the strain tensor is defined as a gradient of displacement vector and then the elastic energy of an element is written in a matrix form using a finite element formulation described in Sect. 2.5

$$W_c = \frac{1}{2}\mathbf{u}^T \mathbf{K}\mathbf{u} \quad \text{and} \quad \mathbf{u} = \{\mathbf{u}_1, \mathbf{u}_2, \cdots \mathbf{u}_n\}^T \tag{4.24}$$

Fig. 4.8 Displacement and
strain of a triangular element

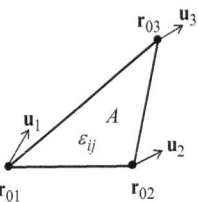

where $\mathbf{K}$ is an element stiffness matrix and $\mathbf{u}_i$ is a displacement vector at node i.
The displacement vector is given by

$$\mathbf{u}_i = \mathbf{r}_i - \mathbf{r}_{0i} \qquad (4.25)$$

where $\mathbf{r}_i$ is a position vector of the node and suffix 0 denotes a reference state.
Therefore, the internal force vectors at nodes are obtained by using the same
formula of (4.6) and are written as

$$\mathbf{f} = -\frac{\partial W_c}{\partial \mathbf{r}} = -\frac{\partial W_c}{\partial \mathbf{u}} = \mathbf{K}\mathbf{u}. \qquad (4.26)$$

Exercise 4.4 Consider a small deformation of the triangular element as shown in
Fig. 4.8. Assuming a linear isotropic material, obtain the x–y components of force
vectors at three nodal points of the element.

4.2   Formulation and Solving Method

4.2.1   Minimum Energy Problem

Consider a deformation of a spring network system consisting of the elements
shown in Sect. 4.1. It is assumed that a body force, $\mathbf{G}_i$, and a surface force, $\mathbf{P}_i$,
applied at nodal point i in the system are conservative forces and respectively
written as

$$\mathbf{G}_i = -\frac{\partial \Phi_G}{\partial \mathbf{u}_i} = -\frac{\partial \Phi_G}{\partial \mathbf{r}_i} \qquad (4.27a)$$

$$\mathbf{P}_i = -\frac{\partial \Phi_P}{\partial \mathbf{u}_i} = -\frac{\partial \Phi_P}{\partial \mathbf{r}_i} \qquad (4.27b)$$

with using potential functions, Φ_G and Φ_P. The total potential energy of the system is

$$\Pi = W + \Phi_G + \Phi_P \qquad (4.28)$$

Fig. 4.9 Motion of a mass
point at node by the force
acting on it

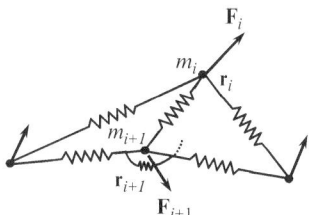

where W is the total elastic energy of all elements ($j = 1 \sim N_e$), including various
elastic components;

$$W = \sum_{j=1}^{N_e} W_j \text{ and } W_j = W_{sj} + W_{aj} + W_{bj} + \ldots . \tag{4.29}$$

Since all the forces are balanced at an equilibrium state, we obtain

$$-\frac{\partial \Pi}{\partial \mathbf{r}_i} = \mathbf{f}_i + \mathbf{P}_i + \mathbf{G}_i = 0. \tag{4.30}$$

This indicates that the total potential energy, Π, is stationary at the equilibrium
state. In physics, Π takes the minimum value at a stable state in equilibrium, which
is called the principle of minimum potential energy.

The above formulation does not alter even when displacement is constrained on
Ω_S. Thus the deformation of the spring network system is described as the minimum
energy problem;

$$\text{Minimize } \Pi \text{ with respect to } \mathbf{r}_i$$
$$\text{subject to } \mathbf{r}_k = \mathbf{R}_k \text{ on } \Omega_S \tag{4.31}$$

which determines the position vectors of nodal points such that the total potential
energy of the system is minimized.

4.2.2 Solving Method

The minimum energy problem of (4.31) can be solved by moving mass points
assigned at nodal points in accordance with the motion equation (Fig. 4.9);

$$m_i \frac{d^2 \mathbf{r}_i}{dt^2} + \gamma \frac{d\mathbf{r}_i}{dt} = \mathbf{F}_i \text{ and } \mathbf{r}_k = \mathbf{R}_k \text{ on } \Omega_S \tag{4.32}$$

Fig. 4.10 Stretch of a rectangular specimen

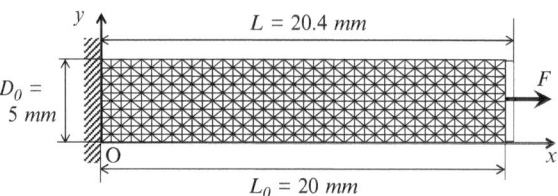

where m_i is a mass and γ is a viscosity. The forces applied to mass points are

$$\mathbf{F}_i = -\frac{\partial \Pi}{\partial \mathbf{r}_i} \qquad (4.33)$$

The motion equation is discretized with a finite difference method and is solved step by step with a sufficiently small time step until a stable state is accomplished. The obtained stationary solution satisfies (4.31) because $\mathbf{F}_i = 0$.

4.3   Parameter Identification of the Spring Network Model

As described in Sect. 1.2, spring constants represent material properties including elasticity and size of the material. This implies that the values of spring constants depend on the mesh size of a discrete model. Although the spring model is easily used in elastic problems in biological systems, it is important to adequately determine spring constants in consideration of the compatibility with a continuum model.

4.3.1   Stretching Spring Constant

Consider a stretch of a rectangular tissue specimen with a width of D_0 and a length of L_0, composed of a linear elastic material. As shown in Fig. 4.10, we define the x–y coordinate system and create a spring network model by dividing the domain into triangular elements ($j = 1 \sim N_e$) and assigning linear springs ($l = 1 \sim N_l$) along edges of elements. Since there are two independent material constants, Young's modulus, E, and Poisson's ratio, ν, for a linear elastic material, one more elastic component is required for a spring network model. Thus, we introduce two sorts of elastic energies, one generated by a stretch of the springs, W_s, and the other generated by a change in the area of elements, W_a;

$$W_s = \frac{1}{2} k_s \sum_{l=1}^{N_l} (L_l - L_{l0})^2 \qquad (4.34)$$

and

$$W_a = \frac{1}{2}k_a\sum_{e=1}^{N_e}\left(\frac{A_e - A_{0e}}{A_{0e}}\right)^2 A_{0e} \qquad (4.35)$$

where suffix 0 denotes a natural state, L_l is the length of the spring l, and A_e is the area of element e. Finally, the problem is described as

$$\text{Minimize } \Pi = W_s + W_a \text{ with respect to } \mathbf{r}_i = (x_i, y_i) \qquad (4.36a)$$

$$\text{subject to } x_i = \begin{cases} 0 \text{ at } x_{0i} = 0 \\ L \text{ at } x_{0i} = L_0 \end{cases}. \qquad (4.36b)$$

Here we assume $D_0 = 5$ mm and $L_0 = 20$ mm, and give a uniaxial strain of $\varepsilon_x = 2\%$ by stretching the specimen in the x direction by $L = 20.4$ mm. The strain energy of a continuum model is written as

$$\Pi = \frac{1}{2}E_\varepsilon^2 V_0 \qquad (4.37)$$

where V_0 is the volume of the tissue specimen at the initial state. Thus the equivalent Young's modulus can be obtained from the total elastic energy of the spring network model by assuming a thickness, t, of the continuum model. Under this condition, the average force applied at both edges of the specimen is identical to that of the continuum model although the force at each nodal point is not uniform in the spring network model. The other material constant, the Poison's ratio, ν, is calculated from

$$\nu = \frac{D_0 - D}{D_0} \qquad (4.38)$$

where D is the width of the stretched specimen.

Figures 4.11 and 4.12 show the relationship between the spring constant, k_s, and the material constants, E and ν, of the corresponding continuum model with a thickness of 1 mm. It is found that the Poisson's ratio is fixed to a single value ($\nu = 0.031$) when the elastic energy of W_a is ignored with $k_a = 0$, but various material constants can be represented by choosing a combination of k_s and k_a. Note that the spring constants depend on the size and mesh structure of a spring network model.

Incompressibility is often assumed for biological materials such as a tissue and a cell. This is represented by taking a large value of k_a so that the area of elements in the 2D spring network model is not changed. Since deformation without area changes is shear, spring constant k_s corresponds to the shear modulus G of the material. It is known that there is a relation, $G = 0.45k_s$, for a random spring network model (Hansen et al. 1996).

Fig. 4.11 Relationship between the equivalent Young's modulus and the spring constant k_s

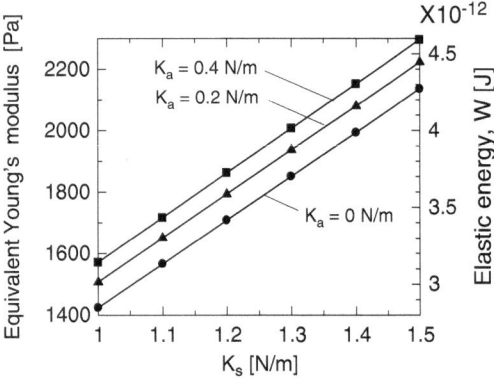

Fig. 4.12 Relationship between the Poisson's ratio and the spring constant k_s

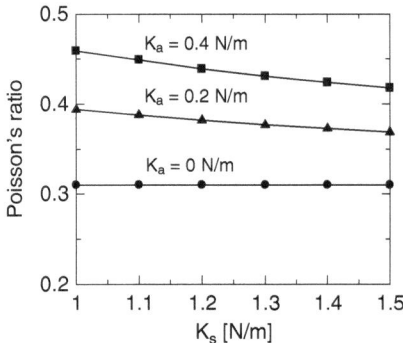

4.3.2 Bending Spring Constant

In a continuum model, bending stiffness is determined from material properties and geometry of the cross-section where a bending moment acts, and its value is dependent on the stiffness of stretching. However, in the 2D spring network model, the bending stiffness is independent of them. Thus, the spring constant of bending, k_b, should be determined to keep a compatibility with the bending stiffness of continuum model.

Here we set up the bending springs ($l = 1 \sim N_l$) between two neighboring elements in the spring network model shown in Fig. 4.10 and make a cantilever by fixing it as $z_i = 0$ and $dz_i/dx = 0$ at $x = 0$. The force P is applied at $x = L_0$, perpendicularly to the x–y plane ($-z$ direction), so that the deflection of the cantilever is $\delta_{\max} = 2$ mm at the tip. The thickness of the cantilever is assumed to be $t = 1$ mm.

Fig. 4.13 Relationship between the bending energy and the spring constant k_b

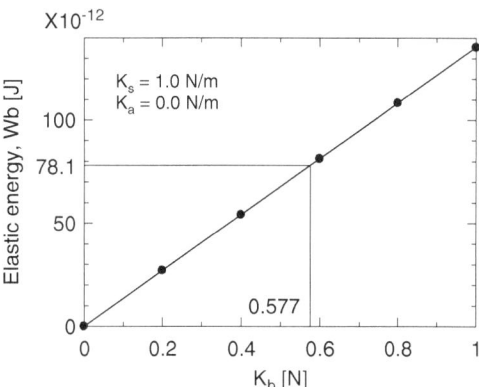

According to the solid mechanics theory, bending energy stored in a cantilever is

$$W = \frac{1}{2} \int_0^{L_0} \frac{M^2}{EI} dx = \frac{P^2 L_0^3}{6EI} \tag{4.39}$$

where M is the bending moment, E is the Young's modulus, and I is the second moment of area; $I = D_0 t^3/12$. On the other hand, bending energy of a spring network model is expressed as

$$W_b = \frac{1}{2} k_b \sum_{l=1}^{N_l} L_l \theta^2. \tag{4.40}$$

Thus, the spring constant k_b can be determined so that $W_b = W$.

Figure 4.13 shows the relationship between the spring constant k_b and the bending energy for $k_s = 1$ N/m and $k_a = 0$ N/m. This combination of spring constants k_s and k_a corresponds to the material constants of $E = 1{,}424$ Pa and $v = 0.31$ (see Figs. 4.11 and 4.12). By substituting these values into (4.39), we obtain the bending energy $W = 78.1 \times 10^{-12}$ J and $k_b = 0.577$ from Fig. 4.13. Figure 4.14 compares the deflection of the cantilever obtained by the spring network model and the continuum model using identified parameters. The difference is less than 1.33% of the maximum deflection.

In the following sections, we introduce some applications of spring network model to biological systems.

Exercise 4.5 Show that the deflection of the cantilever along the axis is given by

$$\delta = \left(1 - \frac{3(L_0 - x)}{2L_0} + \frac{(L_0 - x)^3}{2L_0^3}\right) \delta_{max} \text{ and } \delta_{max} = \frac{PL_0^3}{3EI}. \tag{4.41}$$

Fig. 4.14 Comparison of the deflection of cantilever obtained by a spring network model and a continuum model

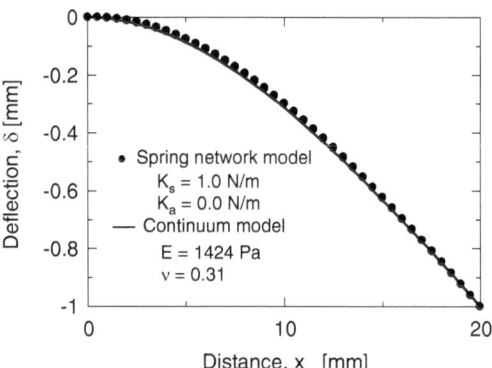

4.4   Mechanical Behavior of a Single Red Blood Cell

4.4.1   Minimum Energy Problem to Determine the Shape of Red Blood Cell

A normal red blood cell (RBC) in blood takes a biconcave discoid shape as shown in Fig. 4.15. It easily deforms in blood flow such that it can pass through a capillary whose diameter is smaller than that of the cell by changing its shape. The shape of RBC also changes under an abnormal condition of blood. Clinically, various shapes of RBC such as spherocyte, stomatocyte, echinocyte, and poikilocyte are observed (Brånemark and Lindström 1963; Brecher and Bessis 1972). Since the shape is determined as an equilibrium state of the mechanical force, the minimum energy concept has been employed to analyze the shape of RBC (Evans 1974; Pai and Weymann 1980; Iglič 1997; Boey et al. 1998; Discher et al. 1998). Canham (1970) explained that the biconcave shape is obtained at the minimum bending energy of the membrane under the constancy of the surface area and volume of the cell. In this section, we express various shapes of RBC using a spring network model.

RBC has no internal structure such as nucleus and actin fibers, and thus its mechanical nature is mostly concerned with that of the membrane (Evans and Skalak 1980). The RBC membrane consists of a lipid bilayer and its underlying meshwork of spectrin. The spectrin is a highly elastic protein fiber and it forms the elastic meshwork which is anchored firmly to the lipid bilayer via various transmembrane proteins.

Since the fluidic bilayer of lipid prevents the area expansion of the membrane, the deformation of the membrane results in shear deformation. Therefore the constitutive law of RBC membrane has been constructed in a continuum model only for the shear deformation under the assumption, $\lambda_1\lambda_2 = 1$, where λ_1 and λ_2 are the principal stretch ratio of the membrane (Skalak et al. 1973). In addition, a bending resistance is also an important mechanical property of the RBC membrane. Those mechanical properties of the membrane can be described by the spring model shown in Sect. 4.1.

Fig. 4.15 Average configuration of RBC (Fung 1993; Tsang 1975)

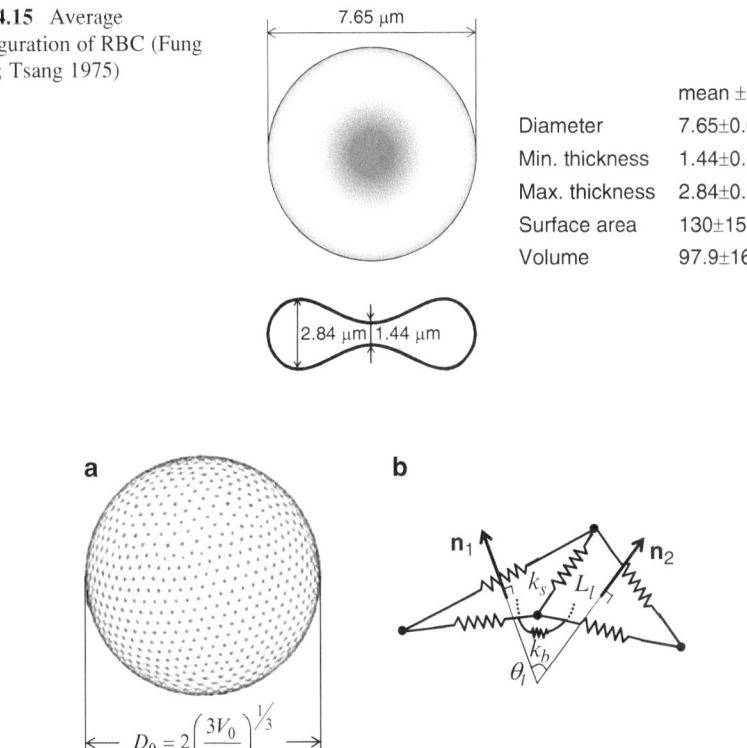

	mean ± SD
Diameter	7.65±0.67 μm
Min. thickness	1.44±0.47 μm
Max. thickness	2.84±0.46 μm
Surface area	130±15.9 μm²
Volume	97.9±16.2 μm³

Fig. 4.16 Spring network model of RBC; (**a**) Initial geometry and (**b**) spring elements of membrane

Here, the initial shape of RBC was assumed to be spherical with the volume of a normal RBC. The membrane was divided into 4,484 triangular elements with a mean edge length of 0.253 μm as shown in Fig. 4.16. The elastic energies for stretching and bending of the membrane are respectively described as

$$W_s = \frac{1}{2}k_s \sum_{l=1}^{N_l} (L_l - L_{l0})^2 \tag{4.42}$$

$$W_b = \frac{1}{2}k_b \sum_{l=1}^{N_l} L_l \tan^2\left(\frac{\theta_l}{2}\right) \tag{4.43}$$

where k_s and k_b are the spring constants, L_l is the length of a spring l, N_l is the number of springs, θ_l is the contacting angle between neighboring elements, and subscript 0 denotes the natural state. The reason why a tangent function is used in the bending energy is to prevent the membrane from folding. There are two kinds of elastic energies in area expansion of the membrane, namely, global and local area changes, which are

$$W_A = \frac{1}{2} k_A \left(\frac{A - A_0}{A_0} \right)^2 A_0 \qquad (4.44)$$

$$W_a = \frac{1}{2} k_a \sum_{e=1}^{N_e} \left(\frac{A_e - A_{0e}}{A_{0e}} \right)^2 A_{0e} \qquad (4.45)$$

where k_A and k_a are constants, A_e is the surface area of the element e, and A is the total surface area of the membrane. If the lipid bilayer can move freely over the spectrin meshwork (Liu et al. 1989), the local area change is not constrained and thus $k_a = 0$. In contrast, if the lipid bilayer moves together with the meshwork, there is no need to consider the global area change and thus $k_A = 0$. Furthermore, the membrane exhibits a pure shear deformation for large k_a, which is represented by a continuity model under $\lambda_1 \lambda_2 = 1$.

Using the elastic energy functions from (4.42) to (4.45), the shape of RBC can be described as (Wada and Kobayashi 2003)

$$\begin{aligned} &\text{Minimize } W \text{ with respect to } \mathbf{r}_i \\ &W = W_s + W_b + W_a + W_A \\ &\text{Subject to } V = V_t \end{aligned} \qquad (4.46)$$

where V is the volume of the RBC and V_t is the desired value. By using a penalty function with the same form of (4.20);

$$W_V = \frac{1}{2} k_V \left(\frac{V - V_t}{V_t} \right)^2 V_t, \qquad (4.47)$$

the volume constraint is included in the potential energy function as

$$\Pi = W + W_V. \qquad (4.48)$$

Then, the problem to determine the shape of RBC is described by

$$\text{Minimize } \Pi \text{ with respect to } \mathbf{r}_i. \qquad (4.49)$$

The material constants of the RBC membrane and the equivalent spring constants are listed in Table 4.1. In the case of the RBC membrane, the area expansion modulus is so large that the deformation results in shear deformation. The spring constant k_s was determined from the shear modulus of membrane.

Table 4.1 Spring constants corresponding to material properties of RBC

Material constants	Spring constants	Remarks
Shear modulus[a] $G = 6\text{–}9 \times 10^{-6}$ N/m	$k_s = 1.3\text{–}2.0 \times 10^{-5}$ N/m	$\mu/k_s = 0.45$[b]
Bending stiffness[c] $B = 0.4\text{–}3 \times 10^{-19}$ J	$k_b = 1.0 \times 10^{-11}$ N[d]	$W_b = W_B$[e]
Area expansion modulus[f] $K = 0.5$ N/m	$k_A = (1-\phi)\,K,\ k_a = \phi\,K$	$0 \le \phi \le 1$

[a]Mohandas and Evans (1994)
[b]Hansen et al. (1996)
[c]Hochmuch and Waugh (1987)
[d]The equivalent spring constant is dependent on the mesh size
[e]$W_B = \frac{1}{2} B \int_\Omega (C_1 + C_2)^2 dA$ $(C_1, C_2 :$ curvature) (4.50)
[f]Evans et al. (1976)

The spring constant k_b was obtained from the bending energy at the initial shape of RBC. The area expansion modulus was separated into k_a and k_A using a weighting coefficient, which was estimated to be $\phi = 0.1$ by comparison of simulations with experimental results of the aspiration test by a micro pipette (Evans et al. 1976). In the actual calculation, the area expansion modulus was too large to ensure the numerical stability and thus it was reduced to one hundredth.

Based on the experimental data (Evans and Fung 1972), the calculation was started from a spherical shape whose surface area was that of a normal RBC ($A_0 = 134$ μm^2) and then the volume ($V_0 = 146$ μm^3) was decreased to that of the normal RBC ($V_t = 87.6$ μm^3). Figures 4.17 and 4.18 show transformation of the RBC shape and the elastic energy while decreasing the RBC volume. The time is normalized by the artificial viscosity γ in the motion equation (4.32) as $t^* = \gamma t$. The shape changes variously with a decrease in the total potential energy, Π, and finally converges to a biconcave discoid shape like a normal RBC at the minimum energy state. It is also found that the stretch (shear) and bending energies dominantly remain at the stable state, indicating that the shape is determined so as to minimize those elastic energies.

In the above simulation, we assumed the RBC membrane with a low shear modulus to obtain a biconcave shape at the final stable state. If the spring constant k_s is increased up to the same order with the identified value shown in Table 4.1, the final shape gets cupped as in Fig. 4.19. By comparison of cupped and biconcave shapes, the average shear deformation is smaller and the average bending deformation is larger for the cupped shape than for the biconcave shape. To minimize the total elastic energy, the shape transforms from biconcave to cupped shape with increasing k_s. This is caused by the assumption of the initial shape of RBC which gives the natural and reference state of the shear deformation of the membrane. In the structural analysis of the biological system, it is important to give adequate material constants as well as the natural and reference state of the deformation (Tsubota and Wada 2010).

Fig. 4.17 Shape
transformation of a swollen
RBC caused by decreasing its
volume

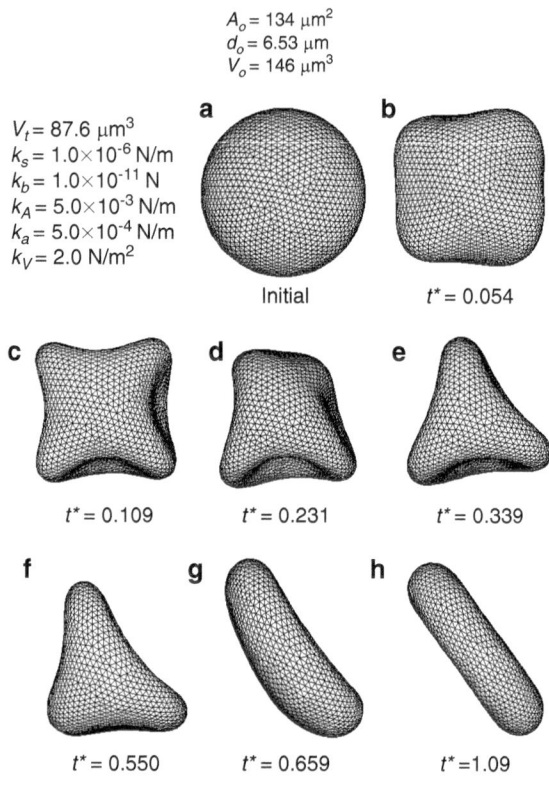

Fig. 4.18 Transition of the
elastic energy of RBC with
change in the shape

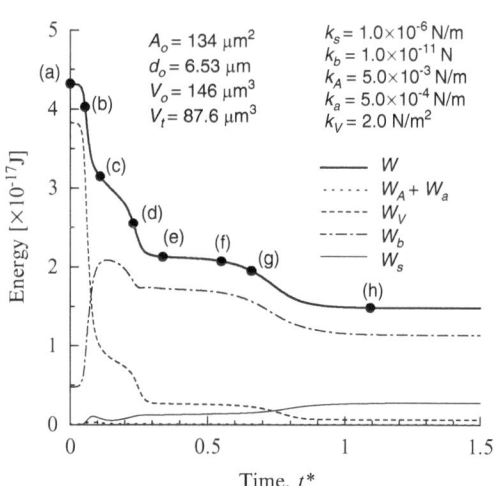

Fig. 4.19 Final stable shape of RBC for different spring constant k_s

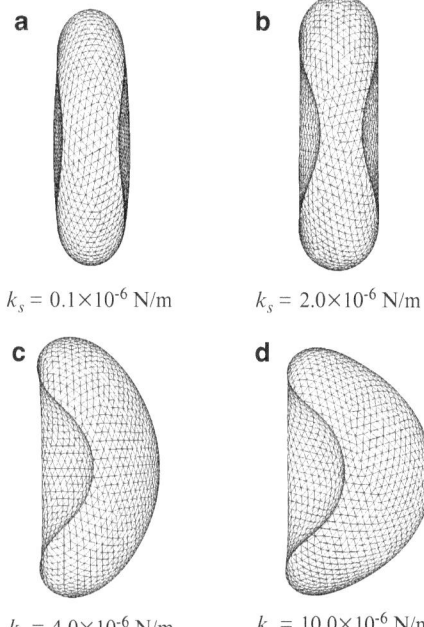

a   b

$k_s = 0.1 \times 10^{-6}$ N/m $k_s = 2.0 \times 10^{-6}$ N/m

c   d

$k_s = 4.0 \times 10^{-6}$ N/m $k_s = 10.0 \times 10^{-6}$ N/m

4.4.2   Red Blood Cell Behavior in a Shear Flow

It is known that the RBC exhibits various behaviors in shear flows because of their elastic nature. First, an RBC elongates and aligns itself at a constant angle to a flow when embedded in shear flows. Second, it may tumble, exhibit a tank-treading motion of the membrane, or both, depending on the shear rate. These behaviors are characteristic properties of RBCs in shear flows and must be expressed well in any RBC models.

Here we expressed the dynamical behavior of RBC in a shear flow by applying fluid forces to the spring network model of RBC in Sect. 4.4.1 as the external force. The fluid forces were modeled in a one-way coupled manner, assuming no influence of the RBC on surrounding flows.

Due to a velocity difference between an RBC and external or internal fluid, fluid forces act on an RBC. An external fluid force $\mathbf{f}^{out}$ is described separately for normal force $\mathbf{f}^{out}_n$ and tangential force $\mathbf{f}^{out}_t$. Based on the Newton's viscosity law and conservation of the fluid momentum, the external fluid forces are defined as

$$\mathbf{f}^{out}_n = \rho \, Q^e \, \Delta \mathbf{u}_n^e \tag{4.51}$$

$$\mathbf{f}^{out}_t = \mu_{out} \Delta \mathbf{u}_t^e / \delta \tag{4.52}$$

where $\Delta\mathbf{u}^e$ is the velocity difference between the external fluid and the element e, ρ and μ_{out} are the density and the viscosity of external fluid, and Q^e is $A_e\Delta\mathbf{u}_n^e$ equal to the rate of flow passing through the element e. In (4.52), δ is the thickness of the equivalent boundary layer estimated from the Stokes' law. The internal fluid force $\mathbf{f}^{in}$ exerted from convection of hemoglobin inside RBC is modeled in a similar way to $\mathbf{f}^{out}$.

Exercise 4.6 According to Stokes' law, the drag force acting on a falling ball in a viscous fluid with very small Reynolds number is given by

$$F_d = 6\pi\mu a v_s \tag{4.53}$$

where μ is the dynamic viscosity of the fluid, and a and v_s are the diameter and the settling velocity of the falling ball, respectively. Show that the thickness of equivalent boundary layer is given as

$$\delta = \frac{4}{9}a \tag{4.54}$$

which satisfies $F_d = |\mathbf{f}^{out}_n + \mathbf{f}^{out}_t|$.

When the RBC model is put in the given velocity field, the position vector, $\mathbf{r}_i$, of node i on the RBC membrane is obtained by solving the motion equation;

$$m_i \frac{d^2\mathbf{r}_i}{dt^2} = \mathbf{f} + \mathbf{f}^{out} + \mathbf{f}^{in} \tag{4.55}$$

where m_i is a virtual mass of nodal point. The elastic force $\mathbf{f}$ is gained by

$$\mathbf{f} = -\partial\Pi/\partial\mathbf{r}_i \tag{4.56}$$

where Π is given by (4.48). Since the external forces include a fluid viscosity, it is not necessary to use the artificial viscosity in (4.32).

Here the viscosities of outside blood and inside hemoglobin were assumed to be $\mu_{out} = 0.003$ Pa s and $\mu_{in} = 0.005$ Pa s, respectively. Furthermore, to express an increase in stretching resistance with elongation of spectrin, a spring constant k_s was defined as a function of stretching ratio λ;

$$k_s = k_{s0}\exp\{\alpha(\lambda - \beta)\} \tag{4.57}$$

where α and β were 2.5 and 1, respectively. The values of α and β were determined by trials and errors by comparing with experimental data (Bessho et al. 2010). The natural state of an RBC was obtained from a swollen spherical shape by reducing its volume by 40% in a quiescent flow (see Sect. 4.4.1).

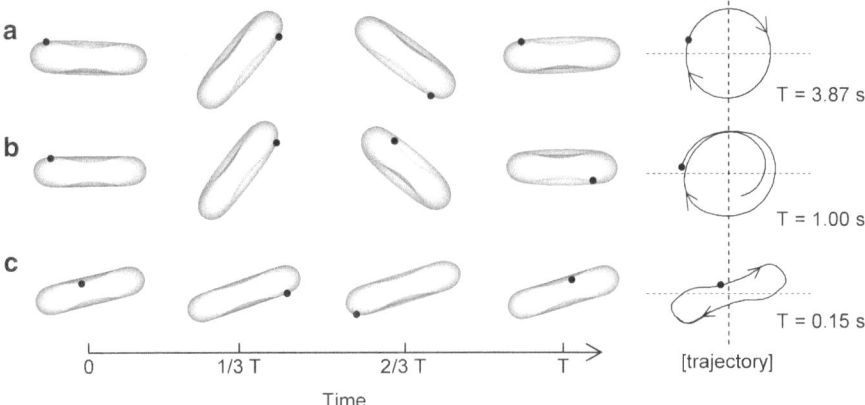

Fig. 4.20   Snapshots of the RBC model subjected to the shear rate (**a**) 5, (**b**) 20 and (**c**) 200 s⁻¹

Figure 4.20 shows a series of snapshots of the RBC model subjected to a shear rate, γ, of (a) 5, (b) 20 and (c) 200 s⁻¹. The membrane was made transparent and marked with a black dot to aid in observing the motion of the RBC. The right figure depicts the trajectory of the black dot. In all conditions, the RBC demonstrated a cyclically repeatable motion. The time required for the cyclic motion was described by cyclic length T. As it can be seen in Fig. 4.20a, the RBC in the flow with a shear rate of 5 s⁻¹ tumbled for $T = 3.87$ s, while keeping a biconcave shape. An increase in the shear rate to 20 s⁻¹ caused a tank-treading motion of the RBC membrane in addition to tumbling, as shown by the non-closed loop of the trajectory. A transition from tumbling to tank-treading occurred at the shear rate of 20–40 s⁻¹. At a higher shear rate, the RBC aligned itself at a constant angle to the flow (around 20°) with the membrane undergoing tank-treading motion. The trajectory at a shear rate of 200 s⁻¹ illustrated in Fig. 4.20c was quite different than that observed when the RBC tumbled.

The frequency of tumbling is plotted against the shear rate in Fig. 4.21. The rate changed in six steps, from 3 to 20 s⁻¹. As it can be seen in the figure, the tumbling frequency increased almost linearly with the shear rate, and their relationship was quantitatively consistent with a theoretical prediction (dashed line) given by Jeffery (1922).

Figure 4.22 plots the tank-tread frequency f of the RBC model as a function of shear rate. For comparison, experimental data from Fischer et al. (1978) is also presented. There was a linear increase in the tank-tread frequency with increasing shear rate, and the f–γ relationship for 11×10^{-3} Pa s was almost the same as that for 18×10^{-3} Pa s.

The dynamic deformation and recovery response of the RBC model was investigated in a cyclically reversing unsteady shear flow at oscillation frequencies, f, of 3 and 5 Hz. Figure 4.23 is a plot of the deformation index L/W (L: long axis,

Fig. 4.21 Plot of the
tumbling frequency against
the shear rate

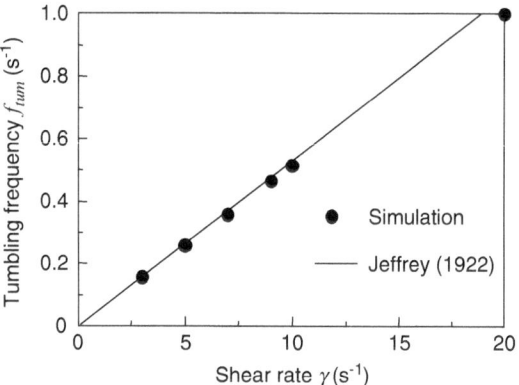

Fig. 4.22 Plot of the tank-
tread frequency against the
shear rate

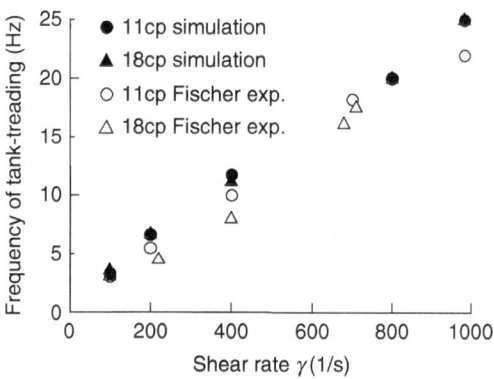

Fig. 4.23 Change in a
deformation index L/W
against shear stress t in an
unsteady shear flow

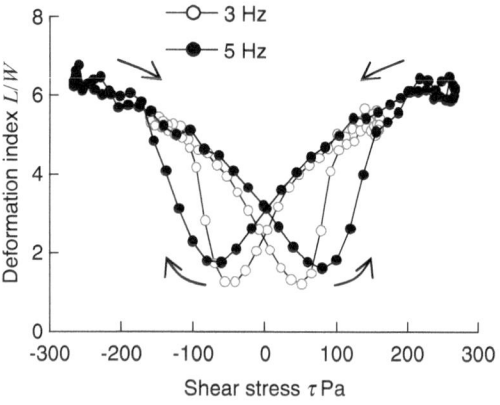

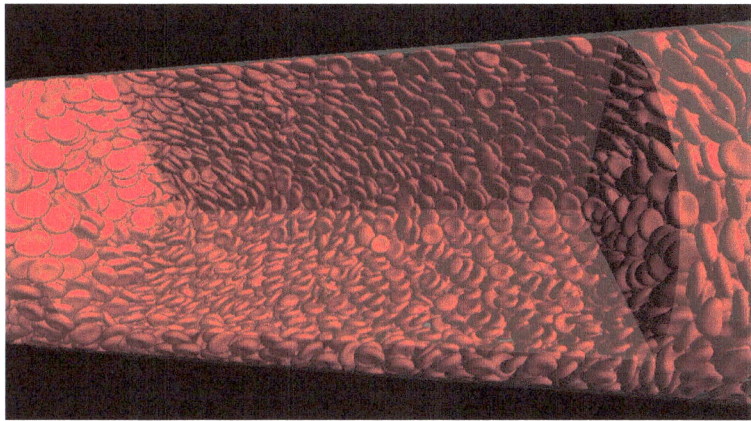

Fig. 4.24 Large-scale simulation of multiple RBC flow. RBC behavior inside the artery is shown by cutting out a quarter cross-section of the blood vessel

W: short axis) against the fluid shear stress τ at an oscillation frequency of 3 Hz. As it can be seen in the figure, the relationship appeared to be a butterfly-like profile regardless of the oscillation frequency. The overall appearance of the curve at 5 Hz was not very different from that at 3 Hz. An increase in the frequency of oscillation extended the loop horizontally and shifted it upwards (Fig. 4.23). The overall appearance of the relationship between the fluid shear stress and the deformation index, L/W, in an oscillatory Couette flow was in good agreement with experimental data (Watanabe et al. 2006). In particular, the maximum of L/W qualitatively agreed, and the horizontal extension and upward shifting of the curve with an increase in the oscillation frequency was also consistent.

The spring network model of RBC shown here may be useful in the analysis of macroscopic phenomena of blood flow, for example, in arterioles, where blood shows both particulate and fluidic properties. At this moment, it is quite challenging even with state-of-the-art supercomputers to handle the behavior of such a number of RBCs and their interactions. Figure 4.24 shows an example of the large-scale simulation of multiple RBCs flow using the present model. Here we put 8,128 RBCs in a straight artery with a diameter of 106 μm and applied a Poiseuille flow with a mean velocity of 500 μm/s. The hematoscrit is 35%. The interaction between RBCs was modeled by a potential function (see Sect. 4.5). The solution domain is divided into 256 volumes for parallel computing.

It would be impossible to solve such a problem with a fully-coupled fluid–structure interaction analysis. The present model does not require solving a matrix equation which saves computational costs by parallelization and enable the large scale simulation. Although further discussion is required for the present RBC flow model, the satisfactory match with experimental results of a single RBC behavior and the possibility of large-scale simulation are encouraging and suggest the potential of such model.

4.5   Mechanical Properties of a Eukaryotic Cell

4.5.1   Mechano-Cell Model

The mechanical properties of eukaryotic cells have been of great interest to scientists from early studies which suggested that mechanical stress-induced alterations in cell shape and structure are critical for controlling many cell functions. Although various loading tests of a cell have been designed to understand the cellular mechanical properties, the heterogeneous intracellular structure such as cytoskeletons brings about difficulties in interpreting experimental data. Computational models have been developed to help such interpretations. Nevertheless, even the tensegrity model (Ingber 2003), which is to date regarded as the most advanced cell model, cannot represent a unified model to explain how mechanical behaviors emerge through collective interactions among different cytoskeletal filaments. This section is aimed to build a mechano-cell model including membrane, cytoskeleton, and nucleus which are the major mechanical components of a eukaryotic cell (Fig. 4.25), and simulate cell deformation and structural changes of cytoskeletons during loading tests.

The mechanical properties of the cell membrane (CM) and nucleus envelope (NE) were modeled in the same manner as that with the RBC membrane shown in Sect. 4.4.1. The CM and NE were represented by a spherical membrane and those surfaces were divided into triangular meshes. Neighboring meshes were connected with a bending spring, while nodes were linked by a spring element to resist to stretching. The total elastic energies of CM and NE are composed of the stretching energy W_s of (4.42), bending energy W_b of (4.33), area expansion energy W_a of (4.44) and W_A of (4.45), and a penalty function W_V of (4.47) for volume constraint and is expressed as

$$W^k = W_s^k + W_b^k + W_A^k + W_V^k \tag{4.58}$$

where k denotes the cell membrane ($k = C$) and the nucleus ($k = N$).

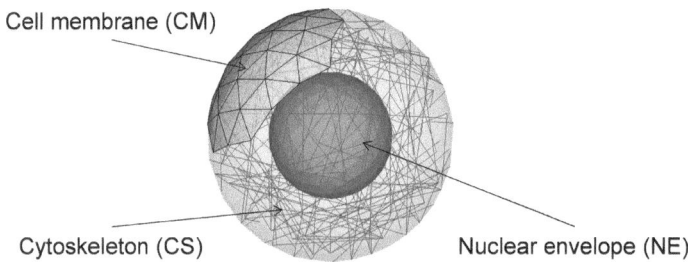

Fig. 4.25 Overview of mechano-cell model

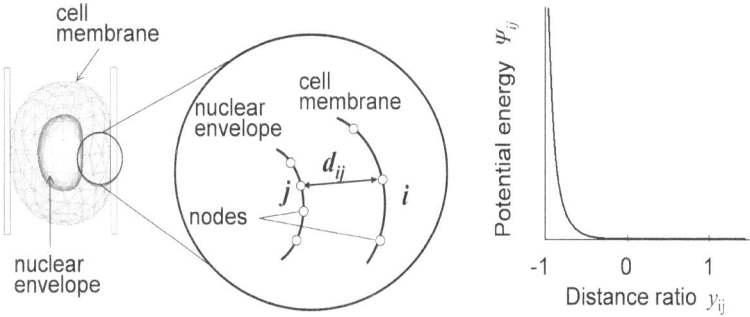

Fig. 4.26 Interaction between cell membrane and nuclear envelope

A cytoskeleton (CS) inside of the cell was modeled by a linear spring which generates a force as a function of its extension. Assuming that CS at the natural state of CM has the natural length l_{0i}, the energy generated by CSs is represented as

$$W^{CS} = k_{CS} \sum_{i=1}^{N_{CS}} (l_i - l_{0i})^2 / 2 \qquad (4.59)$$

where k_{CS} is a spring constant of CS, l_i is a length of CS after deformation, N_{CS} is the total number of CS. Interaction between the CM and NE was represented by a potential function with respect to the distance between the CM and NE;

$$\Psi = \sum_{i}^{N^C} \sum_{j}^{N^N} \Psi_{ij} \qquad (4.60a)$$

$$\Psi_{ij} = \begin{cases} k_n \{ \pi y_{ij}/2 - \tan(\pi y_{ij}/2) \} & \text{for } -1 < y_{ij} \le 0 \\ 0 & \text{for } 0 < y_{ij} \end{cases} \qquad (4.60b)$$

where $y_{ij} = (d_{ij} - \delta)/\delta$, and δ and d_{ij} are the distance between node i on the CM and node j on the NE at the initial state and after deformation (Fig. 4.26).

According to the minimum potential energy principle, the cell deformation is described as

$$\text{Minimize } \Pi \text{ with respect to } \mathbf{r}_i \qquad (4.61a)$$

where

$$\Pi = W^C + W^N + W^{CS} + \Psi. \qquad (4.61b)$$

This was solved by the method shown in Sect. 4.2.2.

Table 4.2 Parameters for cellular membrane and actin fiber as a cytoskeleton

Material constants	Spring constants[a]	Remarks
Elastic modulus of the CM[b]:	$k_s = 5.6 \times 10^{-4}$ N/m	$W_s + W_A = W_D$[d]
$E_{CM} = 1,000$ Pa	$k_A = 2.7 \times 10^{-2}$ N/m	
Poisson's ratio of the CM[c]: $v = 0.3$	$k_a = 3.0 \times 10^{-3}$ N/m	
Bending stiffness of the CM[e]:	$k_b = 9.0 \times 10^{-12}$ N	$W_b = W_B$[f]
$B = 1.0$–2.0×10^{-18} J		
Elastic modulus of the AF[g]:	$k_{AF} = 1.5 \times 10^{-3}$ N/m	$k_{AF} = E_{AF}/S_0 l_0$[h]
$E_{AF} \sim 300$ kPa		

[a]The equivalent spring constant is dependent on the mesh size
[b]McGarry and Prendergast (2004)
[c]Mahaffy et al. (2004)
[d]The strain energy W_D is calculated from (4.25). Thickness of the CM was set to $h = 0.5$ μm (Feneberg et al. 2004)
[e]Zhelev et al. (1994)
[f]From (4.50)
[g]Deguchi et al. (2005)
[h]Cross-sectional area of AF: $S_0 \approx 0.25$ μm², Initial length of AF: $l_0 \approx 10$ μm (Deguchi et al. 2005)

4.5.2   Application of Mechano-Cell Model to Micro Biomechanics

In this section, we demonstrate how the mechano-cell model is useful to understand mechanical properties of a cell in detail, taking a tensile test of a cell as an example.

In a tensile test, a cell was detached from a substrate and floated in an isotonic fluid. Because the shape of such a floating cell is spherical, we assumed that the CM and NE were spheres at their natural state, with diameters of 20 μm and 10 μm, respectively. In the model used, N_s and N_b were both 519, $N_n{}^c$ and $N_n{}^n$ were both 175, N_e was 346. The spring constants estimated from the material properties of a cellular membrane and an actin fiber (AF) are listed in Table 4.2. Since the AFs are mechanically principal components of cytoskeletons, the parameter estimated for actin fibers was used for the cytoskeleton as $k_{CS} = k_{AF}$. There are few data available to estimate the spring constants for the NE, but it has been reported that a nucleus is harder than a cell (Guilak et al. 2000; Caille et al. 2002). Thus, the parameters for NE were assumed to be double those of the CM listed in Table 4.2.

A tensile test was simulated by fixing the nodes at one side while moving those at the opposite side at the speed of 6 μm/s in the direction of cell stretching on the basis of the actual experiment (Ujihara et al. 2010). A load-elongation curve obtained by the simulation is plotted in Fig. 4.27. For comparison, experimental data are also shown as gray lines. The load was assessed as the sum of forces in the stretched direction at the nodes which were actually stretched. As seen, the load increased non-linearly as the cell was stretched.

The cell stiffness S was defined as the slope of the load-deformation curve for every 5 μm deformation D on the basis of the assumption that the curve was

Fig. 4.27 Load-deformation curves obtained from the simulation and the experiments ($n = 10$). The *thick line* represents simulation data, while other *thin lines* with *dots* are experimental data

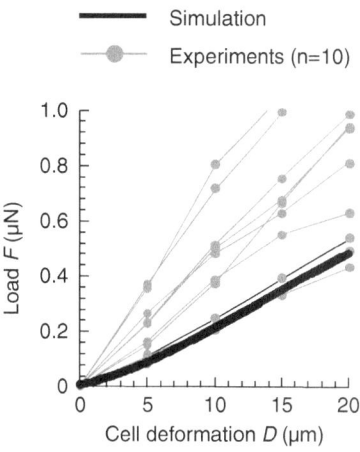

Fig. 4.28 Change in the cell stiffness with cell deformation. The cell stiffness S is evaluated as the slope of the load-deformation curve for a deformation D of every 5 μm, by regression analysis

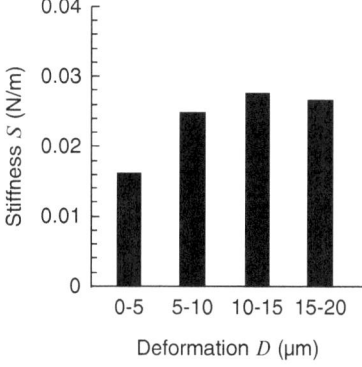

piecewise linear. Figure 4.28 illustrates the cell stiffness S between 0–5, 5–10, 10–15, and 15–20 μm deformation D. The cell stiffness S increased by about 1.5 times as the cell deformation D increased from 0 to 15 μm, whereas it decreased slightly as the cell was stretched further. One may wonder why the cell stiffness increases with cell elongation. This would be clear from Fig. 4.29 showing a deformation behavior of a cell in a tensile test obtained by the simulation and the experiment. The black lines in Fig. 4.29 are AFs. Initially, the cell was spherical and had randomly oriented AFs. Nevertheless, with being stretched, it appeared to be elongated and AFs tended to become aligned and elongated, passively, in the stretched direction, which is visualized in Fig. 4.30 where compressed and stretched AFs are indicated in thick and thin lines, respectively. The passive re-alignment in

Fig. 4.29 Snapshots of a cell during a tensile test (*left*) simulation and (*right*) experiment. A scale is shown at the *bottom* of the figures

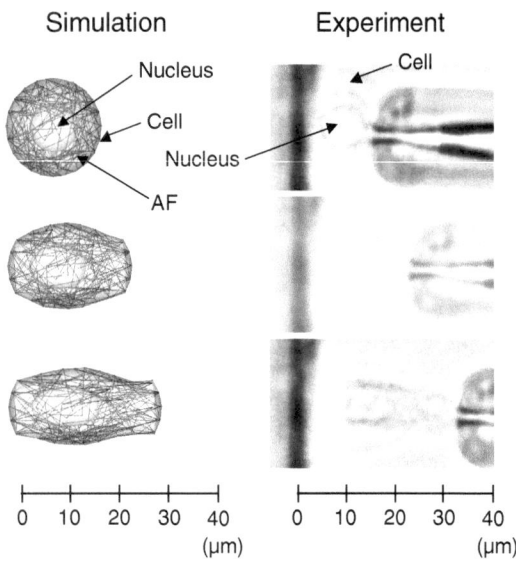

Fig. 4.30 Visualization of AFs during a tensile test. Here, *thick* and *thin* lines represent compressed and stretched AFs, respectively. The compressed AFs are oriented vertically

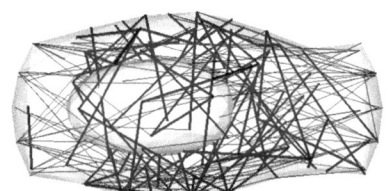

the stretched direction gradually increased the elastic resistance of the whole cell against the stretched direction, causing the load-deformation curve to be non-linear. Therefore, the cell on the whole appears to have a non-linear deformation property although each component of the cell model is expressed by a linear elastic element.

In addition to the tensile test, the mechano-cell model is capable of expressing the cell behaviour in mechanical tests to examine local mechanical properties of a cell such as compression, micropipette aspiration, nano indentation by atomic force microscopy and substrate stretch as exemplified in Fig. 4.31.

Further application of the mechano-cell model is illustrated in Fig. 4.32 where the mechano-cell model is embedded in a tissue. Here, the tissue behavior is described with a continuum mechanics under the assumption of an isotropic linear elastic material, and the behaviors of cytoskeletons within a cell upon the stretch of a tissue are examined. A combined use of the mechano-cell model with the continuum model will help achieve structural integration across physical scales of biomechanical organization from cytoskeletons to tissue.

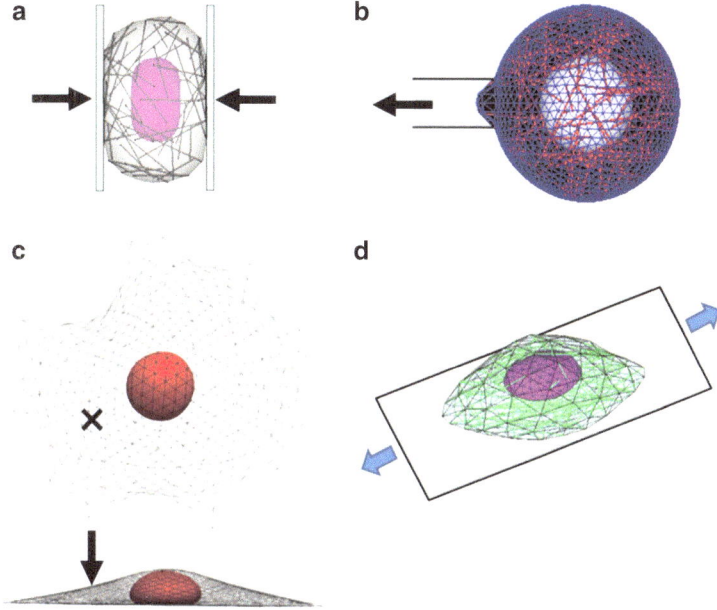

Fig. 4.31 Applications of the mechano-cell model to various mechanical tests; (**a**) compression, (**b**) micropipette aspiration, (**c**) nano indentation, and (**d**) substrate stretch

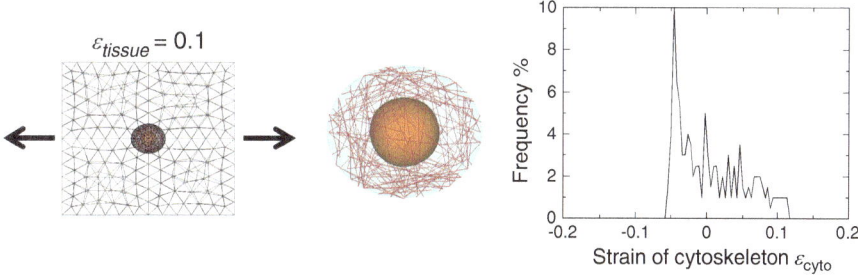

Fig. 4.32 Mechanical behaviors of cell and cytoskeleton in a tissue

4.6 Aneurysm Development

4.6.1 Modeling of Aneurysm

An aneurysm is a vascular disease characterized by local ballooning of an arterial wall. Despite recent progress in the research of aneurysms, the mechanism leading to the initiation and further development of aneurysms remains unclear. If clarified, we expect it becomes possible to forecast the prospects of an aneurysm and make appropriate judgments on its treatment in clinical practice. The etiology of aneurysm formation is thought to be due to interactive changes in mechanical properties of blood vessel wall and consequent vessel geometry (Stehbens 1989; Steiger 1990). In support of this, differences are observed in the expression of certain structural proteins in vessel tissues that represent different phases of the process of aneurysm formation and rupture (Kilic et al. 2005). Furthermore it is obvious that the geometry of an aneurysm remains without blood pressure after the aneurysm is removed. This implies that the permanent deformation which results from growth of the vessel wall is involved in the development of aneurysms (Feng et al. 2005; Shimogonya et al. 2008). In this section we use a hybrid model of a spring network and a continuum tissue to explore how the geometry of a blood vessel changes by partial growth of the vessel wall.

Elastic energy is generated in the vessel wall when it deforms. As energy components, we here considered elastic energies due to in-plane deformation and out-of-plane deformation, i.e. bending. An elastic model based on the continuum mechanics was used to express the in-plane deformation, while a discretized spring model was used for bending. Suppose an artery whose wall is divided into N_e triangular elements, neighbors of which are connected by a bending spring. The total energy generated in the arterial wall by deformation is defined by

$$W = \sum_{j=1}^{N_e} W_c^j + \sum_{k=1}^{N_b} W_b^k \tag{4.62}$$

where W_c^j is the strain energy due to in-plane deformation at element j, W_b^k is the spring energy due to bending at spring k. For a linear elastic material, these energies are given as (4.11) and (4.23), and the spring constant k_b can be determined as described in Sect. 4.3.2.

The problem to determine the vessel geometry is written as

$$\text{Minimize } \Pi \text{ with respect to } \mathbf{r}_i \tag{4.63a}$$

$$\Pi = W + \Phi_F \tag{4.63b}$$

where Φ_F is a potential function of the force generated by a transmural pressure of the blood vessel. The force is given as

$$\mathbf{F}_{\mathbf{P}i} = -\frac{\partial \Phi_F}{\partial \mathbf{r}_i} = \sum_{j=1}^{N_i} \left(\frac{PA_j}{3} \mathbf{n}_j \right) \tag{4.64}$$

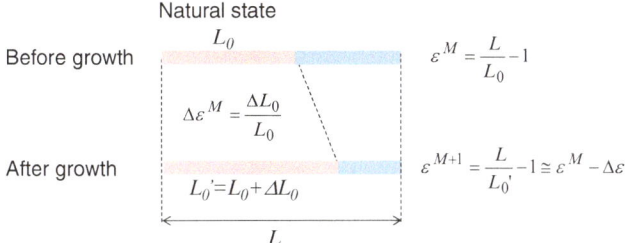

Fig. 4.33 Redefinition of the strain by growth

where $\mathbf{n}_j$ is a unit normal vector to element j, and N_i is the number of elements in which nodal point i is involved. The geometry of the vessel is obtained by solving the motion equation,

$$m_i \frac{d^2\mathbf{r}_i}{dt^2} + \gamma \frac{d\mathbf{r}_i}{dt} = \frac{\partial \Pi}{\partial \mathbf{r}_i} \tag{4.65}$$

until the total potential energy, Π, converges.

Growth of a vessel wall is expressed by updating strains of elements in predefined growing region Ω as

$$\varepsilon_i^{M+1} = \varepsilon_i^M - \Delta\varepsilon_i^M \ (i = x, y) \tag{4.66}$$

when Π converges (Fig. 4.33). Here, ε_i^M is a strain before growth at a convergent time step M, and $\Delta\varepsilon_i^M$ is a strain generated by the growth.

4.6.2   Rule-Based Simulation of Aneurysm Development

The growth of blood vessel during the development of aneurysm has not been elucidated yet. Therefore, we need to incorporate some assumptions of biological events in the theoretical model. We call this the rule-based simulation. The rule assumed in the simulation is verified by comparing the simulated results with the experimental observation. This approach is useful when it is difficult to directly observe biological events (growth of the vessel wall), but possible to observe the resultant behavior (changes in the geometry).

It is known that the mechanical stresses induced by blood flow are involved in the remodeling of arterial wall and the development of vascular diseases such as atherosclerosis and aneurysms (Nerem 1992; Vorp et al. 1998; Burleson and Turitto 1996). Here we assumed that the arterial wall continues to grow isotropically

Fig. 4.34 WSS distribution
in the aorta obtained by CFD
analysis

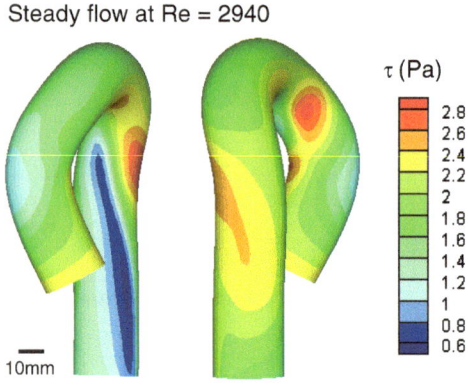

(Δε_x^M = Δε_y^M = Δε^M) at the site where the wall shear stress (WSS), τ, is greater than
a threshold τ_{th}. Based on this assumption, we define the growth strain as

$$\Delta\varepsilon^M = \frac{\tau - \tau_{th}}{\tau_{max} - \tau_{th}}\Delta\varepsilon_{max} \tag{4.67}$$

where τ is WSS, τ_{max} is the maximum value of WSS, and $\Delta\varepsilon_{max}$ is the maximum
strain generated by the growth per a convergence time step. Here we set to τ_{th}
= $0.8\tau_{max}$ and $\Delta\varepsilon_{max}$ = 0.1.

The initial geometry of the blood vessel was made from MR images of a human
aorta that does not have an aneurysm. We ignored the branches stemming from the
aortic arch. Firstly, to determine a high WSS region in the aorta, a mesh model was
generated by dividing lumen of the aorta into hexahedral elements and the CFD
analysis of blood flow was conducted. A finite volume method (see Sect. 3.5.2) was
employed to solve the Navier–Stokes equation and continuity equation. Blood flow
was assumed to be an incompressible Newtonian fluid.

Secondly, for the structure analysis of the blood vessel, the arterial wall model
was constructed by dividing the luminal surface of the aorta into triangular
elements using the nodal points of CFD mesh. The thickness of the wall was
assumed to be constant.

Figure 4.34 shows contour plots of the WSS over the aorta obtained at a steady
flow condition. The Reynolds number is 2,940 at the inlet, which gives a physiologically averaged flow in the aorta. A high WSS was found from the distal end
of ascending aorta to the proximal side of the aortic arch. The maximum WSS
was 2.89 Pa.

Figure 4.35 shows the snapshots of the aorta with the development of aneurysm.
The color denotes the gross of growth strain from the initial state. Saccular
aneurysms were formed at both inside and outside of the aortic arch where WSS
was relatively high at the initial geometry. The configuration of the saccular
aneurysm was similar to clinical observations. Although further discussion is

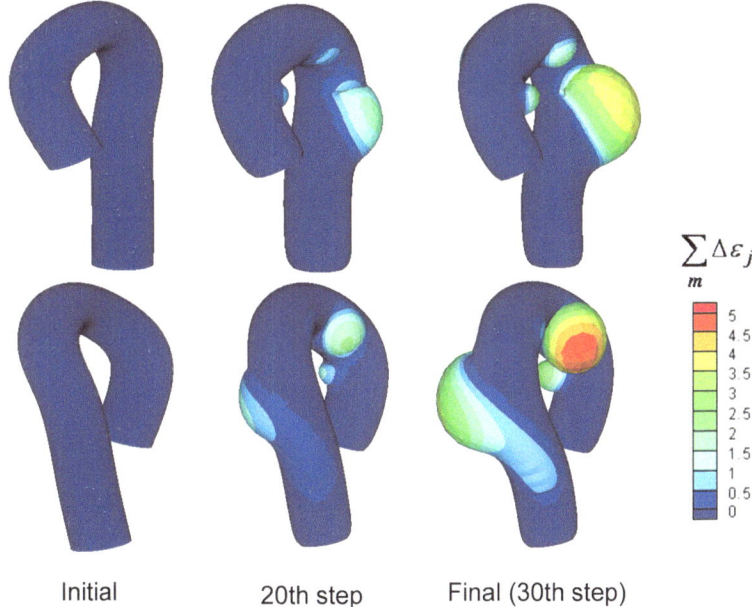

Initial   20th step   Final (30th step)

Fig. 4.35 Development of aneurysms in the aorta

necessary, the rule-based simulations using a spring network model is widely applicable to elucidating hidden mechanisms in biological systems from experimental and theoretical speculations.

4.7   Multi-scale Blood Flow

4.7.1   Modeling of Multiple Red Blood Cell Flow

Blood is a concentrated suspension of blood cells in plasma, an aqueous solution that generally follows Newtonian dynamics. Blood cells are primarily red blood cells (RBCs), comprising about one half of the total blood volume. The particulate nature of RBCs, their inclining and deformability, and physical interactions, contribute significantly to their behavior as a multiphase suspension, which results in a non-Newtonian nature (Secomb 1991). Significant efforts have been made to simulate the rheological behavior of blood by implementing a fluid–structure coupled simulation (Boryczko et al. 2003; Dzwinel et al. 2003; Sun and Munn 2005; Baguchi et al. 2005; Liu and Liu 2006; Zhang et al. 2007, 2009). The fluid–structure-coupled approaches are favorable for analyzing microscopic blood flow, such as a single RBC flowing through a capillary. However, to express mesoscopic blood flow, it is necessary to simulate at least hundreds of RBCs, which

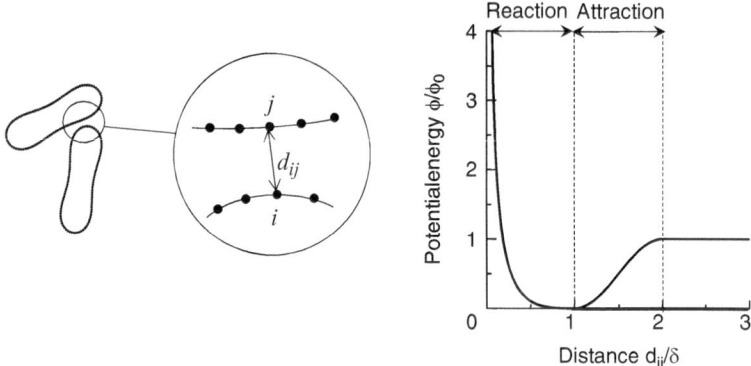

Fig. 4.36 Interaction potential function between two adjacent RBCs

makes the analysis of mesoscopic blood flow quite challenging. Here, we describe a novel computational model to analyze the mesoscopic blood flow where the particulate and continuum natures of blood coexist (Nakamura and Wada 2010).

We modeled blood flow at two different scales, RBC flow at the microscopic level and continuum at the macroscopic level. For the micro-scale flow, a spring network model of RBC and its flow model shown in Sect. 4.4 were used. The interaction of multiple RBCs is described as a potential function,

$$\Phi = \sum_{i=1}^{N}\sum_{j=1}^{N}\Phi_{ij} \tag{4.68a}$$

$$\Phi_{ij} = \begin{cases} k_1(\pi z/2 - \tan \pi z/2) & \text{for } -1 \le z < 0 \\ k_2(1 - \cos \pi z) & \text{for } 0 \le z < 1 \\ 2k_2 & \text{for } 1 \le z \end{cases} \tag{4.68b}$$

where $z = d_{ij}/\delta - 1$, d_{ij} is a distance between the nodal points i and j of nearby two RBCs, and δ is an equilibrium distance (Fig. 4.36).

Flow at a macro-scale was determined by numerically solving continuity and Navier–Stokes equations. The local viscosity of the fluid was modeled as a function of the density of RBC (local hematocrit, Hct) on the basis of Shiga et al. (1979) (see Fig. 4.37 for its conceptual image);

$$\mu_c = \mu_p \exp[k \cdot \text{Hct}] \tag{4.69}$$

where μ_p represents the plasma viscosity and k is a constant, whereby the Navier–Stokes equation is rewritten as

$$\rho(\mathbf{u} \cdot \nabla)\mathbf{u} = -\nabla p + \nabla \cdot \left(\mu_p \exp[k \cdot \text{Hct}]\nabla \mathbf{u}\right) + \mathbf{f} \tag{4.70}$$

which was solved along with the equation of continuity.

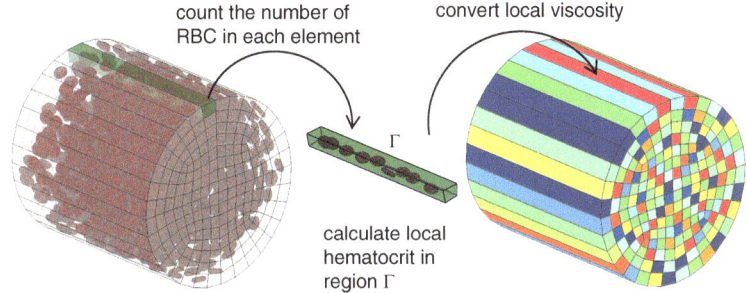

Fig. 4.37 Estimation of the local hematocrit from a microscopic flow simulation and the assignment of local viscosity for the macroscopic flow simulation

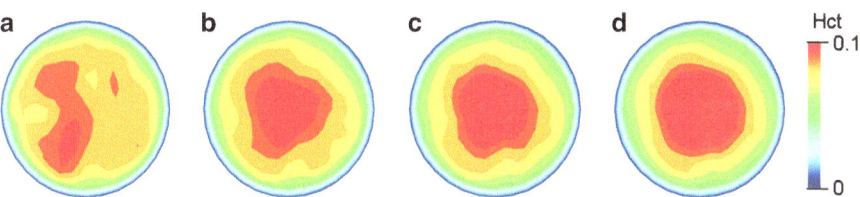

Fig. 4.38 Contour plots of the hematocrit obtained for a global hematocrit of 0.15 at steps (**a**) 1, (**b**) 6, (**c**) 11, and (**d**) 16

4.7.2 Multiscale Simulation of Blood Flow

The calculations of RBC behavior and fluid flow were carried out interactively until a stable flow was obtained. The present study solved hemodynamics in a cylindrical channel having a diameter of 105 μm and a length of 100 μm. In the microscopic simulation, 1,472, 2,416, or 3,417 RBCs were initially placed randomly within the channel, resulting in a global hematocrit of 0.15, 0.24, and 0.35, respectively. The parameter k in (4.69) was set to 2.85. The periodic boundary condition was applied to both edges of the channel for the analysis of RBC flow. The blood flow calculation was carried out for the channel whose axial length was extended to 480 μm. The boundary conditions for the macroscopic simulation included a Poiseuille flow with the Reynolds number of 0.1 at the inlet, a traction-free condition of $p = 0$ at the outlet, and zero velocity with non-slip at the wall.

The results showed a drastic change in the RBC distributions and flow velocities with progress of the simulation. Regardless of the global hematocrit, simulations qualitatively followed the same tendency. As representative simulation results, the spatial distributions of RBCs for Hct 0.15 at steps 1, 6, 11, and 16 are displayed in Fig. 4.38 as a contour plot of hematocrit. And a series of the axial velocity profiles along the y-axis at steps 1, 6, 11, and 16 for Hct 0.15 are presented in Fig. 4.39. Here, the velocity was normalized with the central velocity in the first step.

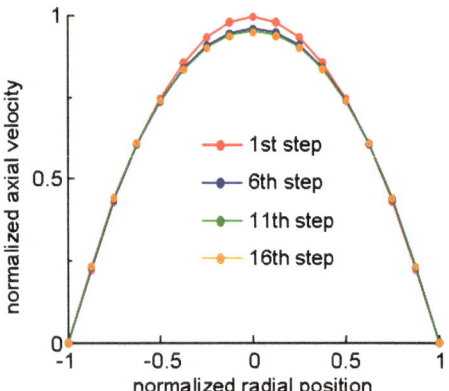

Fig. 4.39 Comparison of the velocity profile of an axial flow obtained for the global hematocrit of 0.15. Note that the velocity was normalized with the central velocity at the first step

Given a Poiseuille flow, RBCs inclined against fluid shear and migrated radially toward the center of flow channel while traveling downstream. Although collisions with other RBCs flowing more inside the channel somehow hindered the migration toward the center of channel, RBCs became more dense at the center of the channel and sparse near the wall, demonstrating a higher hematocrit at the center of the flow channel. Correspondingly, the viscosity became higher at the center of the flow channel. Such a spatial variation in viscosity resulted in a decrease in the velocity of macroscopic blood flow at the center of the channel and an increase near the wall. Also, such a change in the velocity profile of macroscopic flow caused a re-distribution of RBCs within the channel, thereby varying the spatial distribution of a local hematocrit again. An iterative calculation resulted in a progressive decrease in the flow velocity at the center of the channel.

It is quite obvious that RBC behavior induced a change in the macroscopic velocity profile, and vice versa. With the axial migration of RBCs, the RBC concentration became higher around the center of the channel while that near the wall became lower, bringing about an increase in blood viscosity around the center and decreased viscosity near the wall, respectively. As a consequence, the flow velocity in the center of the channel decreased and that near the wall increased, developing into a blunt kind of velocity profile which is observed experimentally (Lima et al. 2007). These results address the potential of the present computational approach in the analysis of the rheology of blood in small vessels where the particulate and continuum natures of blood coexist.

4.8 Summary

This chapter described fundamentals of a spring network model including formulation as the minimum energy problem and identification of the model parameters equivalent to those used in a continuum model. As applications of the spring network

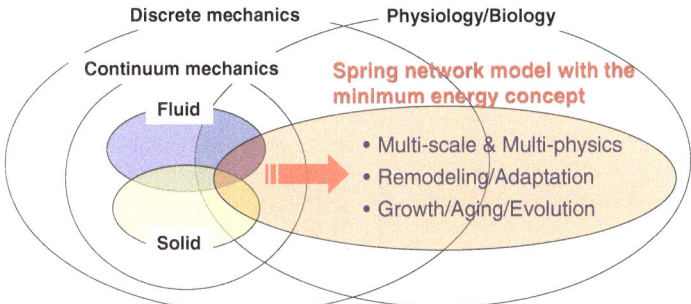

Fig. 4.40 Computational modeling applied to biological systems

model, we described a RBC (red blood cell) model, a mechano-cell model, an aneurysm model and a multiple RBC flow model. All models were described in the same framework of the minimum energy concept even if they included various mechanical components such as line elements, network springs and continuum elements. As demonstrated, the spring network model has a capability to clarify the role and contribution of mechanical components in biological systems, which would help further understanding of interactions between mechanics and biology. As other applications, we described the rule-based simulation of aneurysm development and the multi-scale simulation of blood flow using the hybrid modeling of spring networks and continuum elements. These simulation approaches enable us to explore the possible mechanisms involved in the biological phenomena.

Remodeling, adaptation, growth, aging and evolution are essential of biology. Multi-scale mechanics in molecule, cell, tissue and organ must be a key to understand these biological events. Therefore, we need to extend the mechanics applied to cell biology and molecular biology which include effects of discrete mechanics and molecular dynamics (Fig. 4.40). Spring network modeling with the minimum energy concept is expected to develop the potential of continuum mechanics model by unifying diversely-scaled phenomena in biological systems on the same mechanical platform, whereby giving us an opportunity to further explore physiology and biology from a mechanical standpoint.

References

Baguchi P, Johnson PC, Popel AS (2005) Computational fluid dynamic simulation of aggregation of deformable cells in a shear flow. J Biomech Eng 127:1070–1080

Bessho S, Nakamura M, Wada S (2010) Computational modeling of the behavior of a red blood cell flowing in a high-shear flow (fluids engineering). Trans Jap Soc Mech Eng Ser B 76:2118–2126

Boey SK, Boal DH, Discher DE (1998) Simulations of the erythrocyte cytoskeleton at large deformation. I. Microscopic models. Biophys J 75:1573–1583

Boryczko K, Dzwinel W, Yuen DA (2003) Dynamical clustering of red blood cells in capillary vessels. J Mol Model 9:16–33

Brånemark PI, Lindström J (1963) Shape of circulating blood corpuscles. Biorheology 1:139–147

Brecher G, Bessis M (1972) Present status of spiculed red cells and their relationship to the discocyte-echinocyte transformation. Blood 40:333–344

Burleson AC, Turitto VT (1996) Identification of quantifiable hemodynamic factors in the assessment of cerebral aneurysm behavior. On behalf of the Subcommittee on Biorheology of the Scientific and Standardization Committee of the ISTH. Thromb Haemost 76:118–123

Caille N, Thoumine O, Tardy Y, Meister JJ (2002) Contribution of the nucleus to the mechanical properties of endothelial cells. J Biomech 35:177–187

Canham PB (1970) The minimum energy of bending as a possible explanation of the biconcave shape of the human red blood cell. J Theor Biol 26:61–81

Deguchi S, Ohashi T, Sato M (2005) Evaluation of tension in actin bundle of endothelial cells based on preexisting strain and tensile properties measurements. Mol Cell Biomech 2:125–133

Discher DE, Boal DH, Boey SK (1998) Simulations of the erythrocyte cytoskeleton at large deformation. II. Micropipette aspiration. Biophys J 75:1584–1597

Dzwinel W, Boryczko K, Yuen DA (2003) A discrete-particle model of blood dynamics in capillary vessels. J Colloid Interface Sci 258:163–173

Evans EA (1974) Bending resistance and chemically induced moments in membrane bilayers. Biophys J 14:923–931

Evans EA, Fung YC (1972) Improved measurements of the erythrocyte geometry. Microvasc Res 4:335–347

Evans EA, Skalak R (1980) Mechanics and thermodynamics of biomembranes. CRC, Boca Raton, FL

Evans EA, Waugh R, Melnik L (1976) Elastic area compressibility modulus of red cell membrane. Biophys J 16:585–595

Feneberg W, Aepfelbacher M, Sackmann E (2004) Microviscoelasticity of the apical cell surface of human umbilical vein endothelial cells (HUVEC) within confluent monolayers. Biophys J 87:1338–1350

Feng Y, Wada S, Tsubota K, Yamaguchi T (2005) The application of computer simulation in the genesis and development of intracranial aneurysms. Technol Health Care 13:281–291

Fischer TM, Stöhr-Lissen M, Schmid-Schönbein H (1978) The red cell as a fluid droplet: tank tread-like motion of the human erythrocyte membrane in shear flow. Science 202:894–896

Fung YC (1993) Biomechanics: mechanical properties of living tissue, 2nd edn. Springer, New York, pp 109–164

Guilak F, Tedrow JR, Burgkart R (2000) Viscoelastic properties of the cell nucleus. Biochem Biophys Res Commun 269:781–786

Hansen JC, Skalak R, Chien S, Hoger A (1996) An elastic network model based on the structure of the red blood cell membrane skeleton. Biophys J 70:146–166

Hochmuch R, Waugh RE (1987) Erythrocyte membrane elasticity and viscosity. Annu Rev Physiol 49:209–219

Iglič A (1997) A possible mechanism determining the stability of spiculated red blood cells. J Biomech 30:35–40

Ingber DE (2003) Tensegrity I. Cell structure and hierarchical systems biology. J Cell Sci 116:1157–1173

Jeffery GB (1922) The motion of ellipsoidal particles immersed in a viscous fluid. Proc Roy Soc Lond A 102:161–179

Kilic T, Sohrabifar M, Kurtkaya O, Yildirim O, Elmaci I, Günel M, Pamir MN (2005) Expression of structural proteins and angiogenic factors in normal arterial and unruptured and ruptured aneurysm walls. Neurosurgery 57:997–1007

Lima R, Wada S, Takeda M, Tsubota K, Yamaguchi T (2007) In vitro confocal micro-PIV measurements of blood flow in a square microchannel: the effect of the haematocrit on instantaneous velocity profiles. J Biomech 40:2752–2757

Liu Y, Liu WK (2006) Rheology of red blood cell aggregation by computer simulation. J Comp Phys 220:139–154

Liu SC, Derick LH, Duquette MA, Palek J (1989) Separation of the lipid bilayer from the membrane skeleton during discocyte-echinocyte transformation of human erythrocyte ghosts. Eur J Cell Biol 49:358–365

Mahaffy RE, Park S, Gerde E, Käs J, Shih CK (2004) Quantitative analysis of the viscoelastic properties of thin regions of fibroblasts using atomic force microscopy. Biophys J 86:1777–1793

McGarry JG, Prendergast PJ (2004) A three-dimensional finite element model of an adherent eukaryotic cell. Eur Cell Mater 7:27–34

Mohandas N, Evans E (1994) Mechanical properties of the red cell membrane in relation to molecular structure and genetic defects. Annu Rev Biophys Biomol Struct 23:787–818

Nakamura M, Wada S (2010) Mesoscopic blood flow simulation considering hematocrit-dependent viscosity. J Biomech Sci Eng 5:578–590

Nerem RM (1992) Vascular fluid mechanics, the arterial wall, and atherosclerosis. J Biomech Eng 114:274–282

Pai BK, Weymann HD (1980) Equilibrium shapes of red blood cells in osmotic swelling. J Biomech 13:105–112

Secomb TW (1991) Red blood cell mechanics and capillary blood rheology. Cell Biophys 18:231–251

Shiga T, Maeda N, Suda T, Kon K, Sekiya M, Oka S (1979) Rheological and kinetic dysfunctions of the cholesterol-loaded, human erythrocytes. Biorheology 16(4–5):363–369, No abstract available

Shimogonya Y, Ishikawa T, Imai Y, Mori D, Matsuki N, Yamaguchi T (2008) Formation of saccular cerebral aneurysms may require proliferation of the arterial wall: computational investigation. J Biomech Sci and Eng 3:431–442

Skalak R, Tozeren A, Zarda RP, Chien S (1973) Strain energy function of red blood cell membranes. Biophys J 13:245–264

Stehbens WE (1989) Etiology of intracranial berry aneurysms. J Neurosurg 70:823–831

Steiger HJ (1990) Pathophysiology of development and rupture of cerebral aneurysms. Acta Neurochi Suppl (Wien) 48:1–57

Sun C, Munn LL (2005) Particulate nature of blood determines macroscopic rheology: a 2-D lattice Boltzmann analysis. Biophys J 88:1635–1645

Tsang WCO (1975) The size and shape of human red blood cells. MS Thesis. University of California San Diego, Lajolla, CA

Tsubota K, Wada S (2010) Effect of the natural state of an elastic cellular membrane on tank-treading and tumbling motions of a single red blood cell. Phys Rev E Stat Nonlin Soft Matter Phys 81:011910

Ujihara Y, Nakamura M, Miyazaki H, Wada S (2010) Proposed spring network cell model based on a minimum energy concept. Ann Biomed Eng 38:1530–1538

Vorp DA, Trachtenberg JD, Webster MW (1998) Arterial hemodynamics and wall mechanics. Semin Vasc Surg 11:169–180

Wada S, Kobayashi R (2003) Numerical simulation of various shape changes of a swollen red blood cell by decrease of its volume. Trans Jap Soc Mech Eng Series A 69:14–21

Watanabe N, Kataoka H, Yasuda T, Takatani S (2006) Dynamic deformation and recovery response of a red blood cell to cyclically reversing shear flow: effects of frequency of cyclically reversing shear flow and shear stress level. Biophys J 91:1984–1998

Zhang J, Johnson PC, Popel AS (2007) An immersed boundary lattice Boltzmann approach to simulate deformable liquid capsules and its application to microscopic blood flows. Phys Biol 4:285–295

Zhang J, Johnson PC, Popel AS (2009) Effects of erythrocyte deformability and aggregation on the cell free layer and apparent viscosity of microscopic blood flows. Microvasc Res 77:265–272

Zhelev DV, Needham D, Hochmuth RM (1994) Role of the membrane cortex in neutrophil 772 deformation in small pipets. Biophys J 67:696–705

Chapter 5
Toward In Silico Medicine

The role of computational biomechanics is summarized from a viewpoint of biomechanics modeling and analysis, and a couple of advanced problems are described. The first is the concept of model-based diagnosis that combines computational biomechanics analysis with experimental data. The second is the multi-scale modeling and analysis of structures and functions in biomechanics for biomedicine. The third is the subject-/patient-specific modeling supported by medical measurements and back/inverse analysis. These are key components for in silico medicine towards predictive medicine in the future.

Keywords Evidence-based diagnosis • Model-based diagnosis • Model-based measurement • Multi-scale analysis • Patient-specific analysis

5.1 Computational Biomechanics in Medical Engineering

Biomechanics apply principles, design concepts and problem solving skills of mechanical engineering to medicine and biology. Biomechanics is a relatively new field in the sense of modern engineering science. Early medical engineering applications realized in the last century are easily found in biomechanical prostheses including artificial limbs, artificial hearts, dental implants and so on. With the evolution of computer science and technology, computer analyses along with mathematical modeling of various nature systems in science including biomechanics have become an essential means to explain, explore and gain new insight into phenomena which are too complex to give analytical solutions.

Computational biomechanics has a further potential to advance clinical medicine. Owing to an increase in availability of the high-performance computing resources, much of the work in the area of computational biomechanics has been directed toward detailing and analyzing phenomena, which has resulted in

M. Tanaka et al., *Computational Biomechanics*, A First Course in "In Silico Medicine" 3,   181
DOI 10.1007/978-4-431-54073-1_5, © Springer 2012

further contributions of biomechanics to medicine and biology. First, it is capable of interpolating data that is not obtained by measurements. For instance, if anatomical information of a bone is obtained with some imaging modalities, stress state within the bone becomes available by solid mechanics analysis and used for instance for a fracture risk assessment although stress is not possible to be directly measured with any of means. Provided flow conditions at inlet and outlet of a blood vessel, people are able to grasp flow patterns stimulating the vessel wall by fluid mechanics analysis. Second, the computational biomechanics help people understand complex phenomena from its mechanics on the basis of physics. It is usually quite difficult to gain insights into complex phenomena solely on the basis of observations. Third, once physical explanation is given to the phenomena, it is possible to establish theoretically-supported characteristic indices and quantities that represent the phenomenon by utilizing the power of computational analyses. Many diagnostic indices are submitted empirically or statistically, and their mechanism is rarely proven from a viewpoint of physical science. Forth, computational biomechanics would help quantify pathological conditions of a patient, risk evaluations of treatments, drug efficacy and therapeutic performance, since it provides quantitative data. Finally but ultimately, we expect that computational biomechanics will be able to predict progression of diseases, once we fully elucidate the mechanism that elicits pathophysiological phenomena.

5.2 Model-Based Diagnosis

Figure 5.1 illustrates a general framework of diagnosis. It begins with observations and monitoring including interviews and concludes with diagnosis and prescriptions. A modern way of diagnosis evaluates data obtained medical measurements and observations, and differentiate them by referring empirical and statistical rules, as shown in left column of Fig. 5.1. It is evidence-based diagnosis and medicine, and is highly relied on evidences in practice. The evidence-based approach works effectively when information required in decision-making for diagnosis is sufficiently available. Here, the quality and quantity of all the information and evidences are fully dependent on measurements and observations. Nevertheless successful acquisition of information and data from a patient is not always guaranteed. Accordingly, the outcome of this approach is limited by the ability of modalities. Engineers have tackled such problem, and the evolution of measurement devices has contributed to the advancement of clinical medicine. However, when the phenomenon to be diagnosed becomes more complicated, it is difficult to reconstruct the phenomenon based solely on information obtained by measurement.

The software power resulting from tremendous progress of computer technologies, in these days, complements the progress of hardware technology for medical measurement. Through computational modeling and analyses of structures and functions of human bodies, computational biomechanics is expected to make a

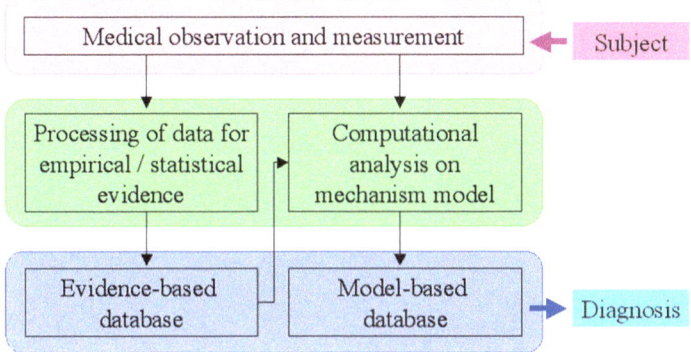

Fig. 5.1 Framework of diagnosis

new epoch in medicine that has so far significantly relied on experimentally and clinically obtained evidences. If the mechanics of a physiological phenomenon occurring inside the human body can be reproduced in a computer, it becomes possible to simulate physiological phenomena, even though not directly observable, based on models of physiological mechanism in conjunction with the measurement. The combined use of computer simulations with measuring devices will contribute to improve the data for diagnosis quantitatively and qualitatively. This is the model-based diagnosis illustrated in the right half of Fig. 5.1.

5.3   Multiscale Modeling and Analysis

Living systems have both physical and biological aspects. The physical aspect is governed by mechanics in major part and can be described within disciplines of mechanics basically. The biological aspects still contain many unknown matters. Biological responses arise from molecular level and their effects are sequentially relayed until they are manifested in cells, in tissues and then in organs. Recent advances in biomechanical science help understand not only the macroscopic mechanical behavior at organ and tissue scales but also microscopic behaviors at cellular and molecular scales. Integration of these multiscale mechanics and biological events is one of the challenging topics in biomechanics. The problems involved in multiscale modeling include the standardization of methodology to collect and present data, the lack of scale linking model that puts phenomena at all scales in one, and public sharing of information.[1] In addition, integration of analyses at different scales in biomechanical problems is impeded by the limited

---

[1] The first volume of this textbook series deals with databasing of models on an open platform for public sharing of information.

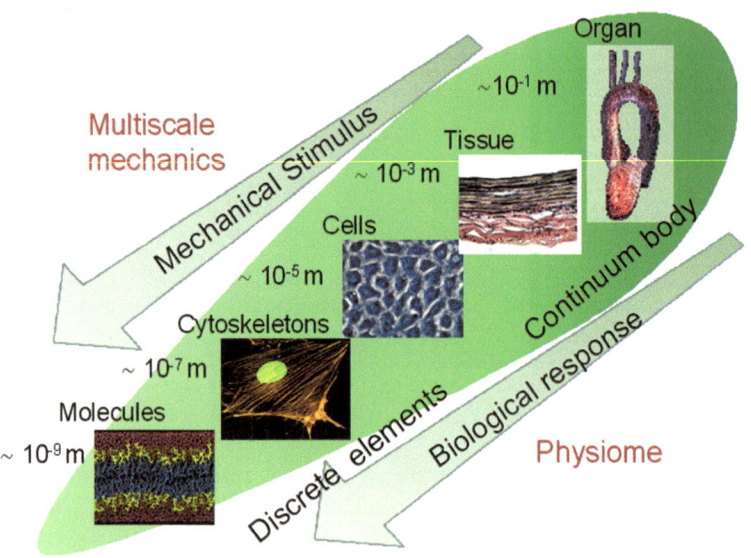

Fig. 5.2 Multiscale framework of structure, mechanical stimuli and biological responses from/to organs to/from cells

amount of information necessary for model development and its verification and validation, the wide variability in anatomical structure and functional properties between individuals, and a highly complicated nature of the underlying physics. In fact, there are still many unknowns in even apparently simple biological and physiological system.

Figure 5.2 illustrates a typical example of the multi-scale problem in biomechanical science. Here, the hierarchical structure of a cardiovascular system is presented. Following levels are usually distinguished: level of organs, level of tissue, level of cell, level of intracellular components such as nucleus and cytoskeletons, and level of molecules including proteins. Particular approaches are used to describe the system at each scale. A continuum model is adopted for describing phenomena from organs to tissues and cells. However, the continuum model may not work properly in the description of behaviors of intracellular elements because at this scale, distribution of elements is spatially discontinuous. Instead, discrete models, or their alternative models are used for this level. For molecular levels, molecular dynamics models are thus far mostly adopted. In fact, researchers involved in the study of each scale level look at a phenomenon with a specific time of length of time, which hinders integration of those models that stand on different mechanical disciplines. Thus one of the critical features is a bidirectional linking between a heterogeneous discrete system at a smaller scale and a homogenized continuum system at a larger scale with different time constants.

Growth, aging, remodeling, adaptation, and evolution are essential in biology. Conventionally, those biological responses are refrained in establishing mechanical models. However, as demonstrated by many biological and physiological studies, biological responses play a pivotal role in maintaining normal functioning of our living systems and constitute adaptation or pathological disease. At this moment, knowledge about biological adaptations and the mechanics associated with this is not sufficient, although attempts have been made to incorporate biological factors such as remodeling, adaptation, growth and deterioration of tissue and cells observed in bone and vascular walls into mechanical models recently with evolution of computational approaches. Thus the modeling of them is another critical feature for multiscale computational biomechanics analyses for long time scale phenomena.

In addition to multi-scalability in time and space, multiphysics modeling is also a big issue in computational modeling of a living system. A heart is totally a multi-physics organ. At a cellular level, polarization and depolarization of cardiac cells is regulated by inflow and outflow of ions of sodium and potassium through ion channels at the cell membrane due to a trans-membrane concentration gradient.[2] Generated action potentials are then propagated through cardiac tissues to cause electrical excitation of a heart. The cardiac muscle then contracts to pump out blood to circulations. Thus, in order to adequately model the biomechanical behavior of the complete heart, even just to satisfy clinically crucial requirements, multiphysics phenomena must be accounted. Moreover, biochemical pathways and cascades must be incorporated in the heart model to characterize a possible pathology, with important perspectives in diagnosis assistance and for therapy planning.

5.4 Subject-/Patient-Specific Modeling and Simulation

Biomechanics problems given in this book are mainly direct analysis of unknown mechanical behavior of biological solids and fluids, and cells. Another analyses mentioned are back or inverse analyses identifying the fundamental data assumed in the direct analysis and adaptive response analyses accompanying the alteration of the target body to be analyzed. Modeling is the essential issue of computational analysis, but is not a specific feature of the computational analysis. Even in very primitive spring- or load-cell-based weight scale for mass measurement, the real quantity measured is the spring elongation or load-cell strain, and a mechanical modeling of a force versus elongation or strain gives us the gravity force and then transformed to the mass as a scale value. Computational mechanics modeling presented in this book works as such a model for the measurement of physical quantities calculated based on the measured values of relevant physical quantities. This kind of analysis is of an inverse or back analysis.

Individual variations of living systems are a kind of obstacles to evaluate biological phenomena from an engineering point of view. Statistical analyses are conventionally applied to process those individual variations in data. Nevertheless,

---

[2] Computational analysis on ion and channel dynamics is extensively described in the first and second volumes of this textbook series.

such individual variations are the essences of problems to be determined specifically for each individual. A subject- or patient-specific modeling arises for those clinical demands. Recent advancements in imaging modalities such as MRI and X-ray CT of course *in vivo* allow us to reconstruct individual geometries of the target organs in silico. For bones, information on their mineral densities can be obtained from CT scans via a calibration with scan data of a phantom. For cardiovascular systems, flow information can be acquired from ultrasound measurements and phase-contrast MRI. Manipulation of such information by computational methods presented in this book combined with *in vivo* experimentation is the first key of biomechanics analysis for practical problems in medicine. This capability can provide a comprehensive understanding on the pathological situation of a specific patient which is unavailable at this moment.

Some fundamental data needed for the subject-specific modeling may be not available in many cases in practice due to limitation of *in vivo* measurements. For instance, the bone mineral density and the deformation of arterial wall are available through medical imaging modalities, but their material properties of elastic modulus are not available within *in vivo* observation only, in spite of being indispensable for mechanical analyses for stress and deformation. Conventionally, these are compensated by using non-subject-specific statistical data. This way of modeling is a semi-subject-specific modeling. However, towards real subject- or patient-specific analyses, an inverse analysis for identification is inevitably required combined with experimental measurement prior to the direct analysis for unknown data of the main target. Experimentation should provide sufficient amount of additional set of information for the data that is inevitable but can not be directly measured, so that the inverse problem becomes well-defined. As is mentioned above, changes caused by adaptation, aging and others are intrinsic of a living body. The meaning of its mathematical modeling is completely different from that for traditional mechanical responses. Nevertheless the modeling concept is the same and it follows the identification problem of a back analysis based on the computational modeling. It is noted that back/inverse analyses have a category of design problem other than that of identification problem. This is ordinarily encountered in the computer-aided engineering analysis and design of common industry problems, and is also found in biomechanics problems that includes an optimal treatment for individual patient by comparing different candidates of treatment through their effects simulated based on a patient-specific modeling. Therefore, a back/inverse analysis is the promising second key as expected for the subject-specific model-based prediction.

5.5   Towards Predictive Medicine

The integration of reductionism biology/medicine and advanced engineering and information sciences is now moving the life science towards a new generation where physiological and pathological information from the living human body can be quantitatively described *in silico* across multiple scales of time and size and through diverse hierarchies of organization—from molecules to cells and organs to individuals. It will have the capacity to develop solutions based upon prior understanding of the dynamic mechanisms and the quantitative logic of human physiology. It will be sure that the establishment of the patient-specific simulation *in silico* will open the gate to a new paradigm "the predictive medicine", probably as a third tool of clinical medicine in addition with medical imaging and experience. The predictive medicine is expected to provide information us to make diagnosis with consideration of how this disease will be in future if not treated or treated in different ways. Once all uncertainties in physiology of an interest are adequately controlled, the computational model can become "predictive", and hence should provide clinically important information, both in the current state of the patient and under various scenarios of future evolutions. The computational modeling, analysis and simulation play the key role in all of these aspects towards *in silico* medicine.

Index